Road Maintenance Management

Other titles of related interest

Building Procurement, Second Edition   Alan Turner
Civil Engineering Contract Administration and Control, Second Edition   I. H. Seeley
Civil Engineering Materials, Fourth Edition   N. Jackson and R. Dhir
Civil Engineering Quantities, Fifth Edition   I. H. Seeley
Constructability in Building and Engineering Projects   A. Griffith and A. C. Sidwell
Finite Elements – A Gentle Introduction   David Henwood and Javier Bonet
Fundamental Structural Analysis   W. J. Spencer
Highway Traffic Analysis and Design, Third Edition   R. J. Salter and N. Hounsell
Introduction to Global Financial Markets   S. Valdez
Management Control   A. J. Berry, J. Broadbent and D. Otley
Plastic Analysis of Steel and Concrete Structures, Second Edition   S. J. J. Moy
Reinforced Concrete Design, Fourth Edition   W. H. Mosley and J. H. Bungey
Reinforced Concrete Design to Eurocode 2   W. H. Mosley, R. Hulse and J. H. Bungey
Soil Mechanics – Principles and Practice   G. Barnes
Structural Mechanics   J. Cain and R. Hulse
Surveying for Engineers, Third Edition   J. Uren and W. F. Price
Understanding Hydraulics   L. Hamill
Understanding Structures, Second Edition   Derek Seward
Urban Land Economics, Fourth Edition   Jack Harvey
Value Management in Construction   B. R. Norton and W. C. McElligott

WA 1159876 X

Road Maintenance Management

Concepts and Systems

Richard Robinson, Uno Danielson and Martin Snaith

MACMILLAN

© The University of Birmingham and The Swedish National Road Administration 1998

All rights reserved. No reproduction, copy or transmission of this publication may be made without written permission.

No paragraph of this publication may be reproduced, copied or transmitted save with written permission or in accordance with the provisions of the Copyright, Designs and Patents Act 1988, or under the terms of any licence permitting limited copying issued by the Copyright Licensing Agency, 90 Tottenham Court Road, London W1P 9HE.

Any person who does any unauthorised act in relation to this publication may be liable to criminal prosecution and civil claims for damages.

The authors have asserted their right to be identified as the authors of this work in accordance with the Copyright, Designs and Patents Act 1988.

First published 1998 by
MACMILLAN PRESS LTD
Houndmills, Basingstoke, Hampshire RG21 6XS
and London
Companies and representatives throughout the world

ISBN 0-333-72155-1

A catalogue record for this book is available from the British Library.

This book is printed on paper suitable for recycling and made from fully managed and sustained forest sources.

10 9 8 7 6 5 4 3 2 1
07 06 05 04 03 02 01 00 99 98

Printed in Great Britain by
Antony Rowe Ltd
Chippenham, Wiltshire

1159876 X

Learning Resources
Centre

Contents

Foreword   xii
Preface   xiv
Acknowledgements   xvi

1 Management Issues   1

    1.1 Purpose of management   1
       1.1.1 Roads as an asset   1
       1.1.2 Activities   2
       1.1.3 Impacts   3
    1.2 Construction and maintenance   4
       1.2.1 Projects and processes   5
       1.2.2 Purpose of maintenance   7
       1.2.3 Reducing deterioration   8
       1.2.4 Lowering vehicle operating costs   10
       1.2.5 Keeping the road open   10
       1.2.6 Safety   11
       1.2.7 Environmental issues   11
    1.3 The management process   11
       1.3.1 Management functions   13
       1.3.2 Project management   14
       1.3.3 The management cycle   17
       1.3.4 Cycles within the management functions   17
    1.4 Quality management   17
       1.4.1 The concept of customers   17
       1.4.2 Quality management systems   20
       1.4.3 Risk management   22
    1.5 Management information   23

2 Institutional Issues   25

    2.1 Management in context   25
       2.1.1 External and internal factors   25
       2.1.2 Economic factors   26
       2.1.3 Roles and aims   27
       2.1.4 Stakeholders   29
    2.2 Legal ownership   29
       2.2.1 The concept of ownership   29
       2.2.2 Ownership and management of capital assets   30

	2.3	The policy framework		31
		2.3.1	Management and policy	31
		2.3.2	Components of the policy framework	32
		2.3.3	Vested interests	33
		2.3.4	Mission statement	34
		2.3.5	Objectives	36
		2.3.6	Standards and intervention levels	36
		2.3.7	Functional hierarchy	38
		2.3.8	Basis for decision-making	39
		2.3.9	Performance indicators	40
	2.4	Organisations		43
		2.4.1	Organisational culture	43
		2.4.2	Size and degree of decentralisation	46
		2.4.3	Organisational status	49
	2.5	Effectiveness and efficiency		50
		2.5.1	Basic concepts	50
		2.5.2	Specificity	50
		2.5.3	Competition	52
		2.5.4	Functional separation	53
		2.5.5	Commercialisation	55
	2.6	Institutional development		56
		2.6.1	Hierarchy of management issues	56
		2.6.2	The challenge of change	56
	2.7	Human resource management		58
		2.7.1	Human resource needs in a changing environment	58
		2.7.2	Skills and motivation	58
		2.7.3	Training	59
	2.8	Environmental issues		61
		2.8.1	Issues of concern	61
		2.8.2	Environmental assessment	61

3 Finance   64

	3.1	Issues		64
	3.2	Allocation of funds		65
		3.2.1	Budgetary process	65
		3.2.2	Budget heads	66
		3.2.3	Allocation mechanisms	67
		3.2.4	Disbursement mechanisms	68
	3.3	Revenue		69
		3.3.1	Basic aims	69
		3.3.2	Principles of pricing and cost recovery	69
		3.3.3	Sources of revenue	71
		3.3.4	Financing road maintenance and development	72
		3.3.5	Managing the revenues	74

3.4	Road funds		74
3.5	Congestion pricing		76
	3.5.1	Managing congestion	76
	3.5.2	Indirect charging mechanisms	77
	3.5.3	Direct charging mechanisms	78
3.6	Road tolls		79
	3.6.1	Advantages	79
	3.6.2	Disadvantages	80
	3.6.3	Pricing the toll	80
	3.6.4	Investment criteria	81
	3.6.5	Design criteria	82
3.7	Private financing		83
	3.7.1	Background	83
	3.7.2	Advantages	85
	3.7.3	Disadvantages	85
	3.7.4	The parties involved	86
	3.7.5	Contractual framework and risks	87
	3.7.6	Shadow tolls	88
3.8	Local financing		90

4 Benefits and Costs   93

4.1	Basis for management decisions		93
4.2	Benefits		93
	4.2.1	Impacts and beneficiaries	93
	4.2.2	Project level benefits	95
	4.2.3	Network level benefits	95
4.3	Costs		97
	4.3.1	Issues to be considered	97
	4.3.2	Methods of accounting	98
	4.3.3	Cost estimating techniques	99
	4.3.4	Global techniques	101
	4.3.5	Unit rate techniques	102
	4.3.6	Operational costing techniques	103
	4.3.7	Worker hours techniques	110
	4.3.8	Suitability of individual techniques	110
	4.3.9	Application of the approach to cost estimating	112
4.4	Cost–benefit analysis		112
	4.4.1	Purpose of the analysis	113
	4.4.2	Prices	113
	4.4.3	Comparison of alternatives	114
4.5	Life cycle costing		117
	4.5.1	Investment choice	117
	4.5.2	Road administration costs	117

		4.5.3 Road user costs	118
		4.5.4 Life cycle cost models	118

5 Management Information — 121

	5.1	Information and data	121
		5.1.1 The need for information	122
		5.1.2 Information groups	122
	5.2	Data design	122
		5.2.1 Data acquisition costs	123
		5.2.2 Relevance	124
		5.2.3 Appropriateness	125
		5.2.4 Reliability	125
		5.2.5 Affordability	126
		5.2.6 Information quality	129
	5.3	Strategies for data collection	129
		5.3.1 Options available	129
		5.3.2 Cost–benefit analysis	130
	5.4	Inventory data	130
		5.4.1 Types of road inventory	130
		5.4.2 Network referencing	130
		5.4.3 Item inventory	132
	5.5	Traffic data	136
		5.5.1 Types of traffic information	136
		5.5.2 Traffic characteristics and volume	136
		5.5.3 Axle loading	141
	5.6	Pavement data	142
		5.6.1 Defects	142
		5.6.2 Manual methods of pavement assessment	145
		5.6.3 Mechanised methods of pavement assessment	146
	5.7	Data management	149
		5.7.1 Computer-based data management	149
		5.7.2 System selection and procurement	153
		5.7.3 Requirements	155
		5.7.4 Specification	155
		5.7.5 Procurement	157
		5.7.6 Implementation	160

6 Treatment Selection — 161

	6.1	Scheduled and condition-responsive methods	161
		6.1.1 Approaches to treatment selection	161
		6.1.2 Scheduled	161
		6.1.3 Condition-responsive methods	163
		6.1.4 Methods available	163

	6.2	Common features	164
		6.2.1 Methods of defect assessment	164
		6.2.2 Precedence rules	164
		6.2.3 Sectioning	165
	6.3	Defect-based rules	166
	6.4	Rules based on condition indices	166
	6.5	Complex rules approaches	173
		6.5.1 Non-transparent methods	173
		6.5.2 Expert systems	174
	6.6	Optimisation approach	177
	6.7	Issues arising from condition-responsive methods	178
	6.8	Treatment selection for gravel roads	179
		6.8.1 Incidence of gravel roads	179
		6.8.2 Defects and deterioration	179
		6.8.3 Treatment selection and design	182
		6.8.4 Assessment of need	183
	6.9	Contract packaging	186
7	**Prioritisation**		**188**
	7.1	Basic concepts	188
		7.1.1 Generations of decision-support systems	188
		7.1.2 Commercial and user models	189
	7.2	First generation prioritisation methods	190
		7.2.1 Basic principles of the methods	190
		7.2.2 Methods based on defectiveness indices	191
		7.2.3 Methods based on degree of defectiveness	191
		7.2.4 Treatment-based methods	194
	7.3	Second generation methods	196
		7.3.1 Basic principles of the methods	196
		7.3.2 Cost-effectiveness methods	196
	7.4	Third generation methods	200
		7.4.1 Statement of the problem	200
		7.4.2 Objective function	200
		7.4.3 Total enumeration	201
		7.4.4 Dynamic programming	201
		7.4.5 Economic boundary	202
		7.4.6 Year-by-year analysis	206
		7.4.7 Consideration of uncertainty	206
	7.5	Reduced time period analysis methods	207
		7.5.1 Theoretical concerns over life cycle analysis	207
		7.5.2 Basis of the approaches	208
		7.5.3 The budget period method	208
		7.5.4 Annualised cost method	209

7.6		Prediction of deterioration	210
	7.6.1	Approaches to condition projection	210
	7.6.2	Probabilistic methods	211
	7.6.3	Deterministic methods using mechanistic models	213
	7.6.4	Deterministic methods using regression models	214
	7.6.5	Deterministic methods using mechanistic–empirical models	214
	7.6.6	Calibration of deterministic methods to fit local conditions	216
7.7		Observations on prioritisation methods	218
	7.7.1	Key features of available methods	218
	7.7.2	Data requirements	219
	7.7.3	Use of third generation methods	219

8 Operations Management — 221

8.1		Operations considered	221
8.2		Routine maintenance	222
	8.2.1	Management tasks and work types	222
	8.2.2	Cyclic works	222
	8.2.3	Reactive works	224
	8.2.4	Scheduling	225
	8.2.5	Utility works	226
	8.2.6	Key considerations	227
8.3		Winter maintenance	227
	8.3.1	The winter problem	227
	8.3.2	The winter maintenance management cycle	229
	8.3.3	Defining aims	230
	8.3.4	Assessing needs	230
	8.3.5	Determining actions	235
	8.3.6	Determining costs and priorities	235
8.4		Choice of technology	237
	8.4.1	Technology options	237
	8.4.2	Technological appropriateness	238
	8.4.3	Cost-effectiveness	238
	8.4.4	Availability of labour and equipment	240
	8.4.5	Domestic resources	241
	8.4.6	Discussion on choice of technology	241
8.5		Equipment management	241
	8.5.1	Basic principles	241
	8.5.2	Autonomous management	242
	8.5.3	Organisation of maintenance and repair	243
	8.5.4	Administrative and commercial management	244
8.6		Monitoring	246
	8.6.1	Issues to be considered	246

		8.6.2	Key indicators of effectiveness	246
		8.6.3	Value for money	247
		8.6.4	Approach to monitoring	247

9 Procurement and Contracts — 248

    9.1 Procurement options 248
        9.1.1 Options available 248
        9.1.2 Competitive procurement 248
        9.1.3 Contracting out construction and periodic maintenance works 249
        9.1.4 Contracting out routine maintenance and emergency works 251
        9.1.5 Contracting out the client functions 253
        9.1.6 Considerations for maintenance by contract 256
    9.2 General procurement issues 263
        9.2.1 Basic principles 263
        9.2.2 Procedures and forms of contract 263
        9.2.3 Qualification 267
        9.2.4 Bidding documents and contracts 268
    9.3 Contract management 269
        9.3.1 Management principles 269
        9.3.2 Works supervision 271
        9.3.3 Quality management 272
    9.4 Contractor development 272
        9.4.1 Basis of the approach 272
        9.4.2 Preliminary study 272
        9.4.3 Implementing the development programme 273
    9.5 Contracts for in-house works 273
        9.5.1 Improvement of existing operations 273
        9.5.2 Competitive procurement of activities 274
    9.6 General observations 277

Appendix A: Management Summary   279

References   280

Index   289

Foreword

On Behalf of the World Road Association (PIARC)

By the Chairman of PIARC Committee on Road Management, Gary Norwell

Road Administrations are responsible for the management of an extremely valuable public asset. They are expected to invest maintenance funding in a way that returns maximum benefit to road users and the communities which they serve. This book recognises this principle and promotes the need to maintain the asset to a standard that will optimise road user costs and the cost of maintenance. This concept is fundamental to the national economy, the pursuit of international competitiveness and the achievement of ecologically sustainable transport.

The book shows road maintenance management as a continuous process which allows all of those involved to see how their role fits into the overall context of managing the asset. It also highlights the value of implementing decision support systems and the need to involve a range of disciplines including engineering, safety, economics, finance, commerce, environment and management.

Road asset managers will also find the book valuable in developing a case for an adequate level of maintenance funds, giving them the ability to demonstate that the timely and effective instrument of asset preservation funds will result in long-term benefits that far exceed the maintenance investment.

The World Road Association (PIARC) Committee on Road Management has identified a wide range of issues that are of interest to member countries and I am pleased to see that many of these have been addressed in this book. PIARC is committed to sharing best practice in all areas of road administration and management, and I believe that this book will become a very useful reference for road managers and practitioners in many countries.

World Road Association (PIARC)
La Grande Arche de la Défense,
92055 LA DEFENSE CEDEX,
FRANCE.

On Behalf of The World Bank

By the World Bank Roads Adviser, Ian G. Heggie

Few people realize that roads are big business. The French Road Administration manages assets roughly the same size as General Motors, while the Japan Road Improvement Special Account generates roughly the same revenues as Nippon Steel and

Pepsico. Like any other important economic activity, roads do not need to be managed like a social service – they can be managed like a business. That is the purpose of this timely book. Countries world-wide are restructuring their national road agencies to make them more commercial. This book tells you how to do that. Drawing on recent work on road financing, it also tells you how to put the financing of roads on a sustainable long-term basis. The book should be essential reading for all those interested in the road sector. è

The World Bank Group
1818 H Street N.W.
Washington
DC 20433
USA

Preface

Traditionally, the core of engineering activity in the roads sub-sector has consisted of the design and construction of new roads. Increasingly, however, road networks have been substantially completed in many countries. The focus of attention is now moving away from building new roads to maintaining existing roads. However, road maintenance is a fundamentally different process from that of building new roads. Construction activity usually involves projects with a defined start and finish; conversely, maintenance is continuous. Whereas design and construction are dominated by engineering issues, maintenance is essentially a management problem. The improvement of maintenance often involves institutional reform, human resource development, and changes to management practices before addressing technical issues. In many countries, it has proved to be very difficult for maintenance to be carried out effectively, resulting in rapid deterioration of many roads.

The purpose of this book is to provide an up-to-date description of road maintenance management. This is a field of rapidly growing importance, and one in which many engineers, and managers from other backgrounds, are increasingly becoming involved. The book has been developed based on the experience of project work in this field in a number of countries, and from a series of short courses run for road practitioners at the University of Birmingham in the United Kingdom, and at the Royal Institute of Technology at Stockholm in Sweden. The book is designed as a basic text for short courses aimed at engineers and managers working in road administrations who have the responsibility for managing maintenance of the existing road network and for implementing road management systems. This includes staff in national, regional and district road administrations; staff in consultancy practices and contracting organisations who are working in the field of road maintenance management; and university postgraduate and undergraduate students pursuing courses in road engineering, transport planning and road management. The book is also aimed at staff in international and bilateral aid agencies who have the responsibility of designing and implementing projects for road management and maintenance.

A key tenet of the book is that road maintenance management can be divided conveniently into four main functions, which have been termed *planning, programming, preparation* and *operations*. These functions provide a framework for considering the part of the network being managed, the time-scale being considered, the staff who are most interested in the management decision, the part of the policy framework that is most relevant, the level of data detail that is appropriate, and the appropriate method of cost estimating to be used. The chapters of this book can be considered to fall into three groups. Chapters 1 to 3 deal with fundamental policy aspects of maintenance management, institutions and finance. They cover subjects that will be of primary concern at the national level, where matters of policy, objectives and standards are normally decided. Chapters 4 to 7 are mainly concerned with planning and programming, which are considered from the perspective of the costs and benefits of investing in maintenance, data, treatment selection and prioritisation. These chapters should be of most interest to those tiers of the organisation whose staff are sub-sector budget holders, possibly at regional level. The remainder of the book focuses on preparation and operational issues of day-to-day management of roads and the procurement of works. These chapters should be of most interest to those working at the detailed levels of road maintenance management, possibly in districts or operational organisations.

This book is a reference textbook for the Overseas Road Note entitled *Guidelines of design and operation of road management systems*, which has been prepared by the Transport Research Laboratory.

The book is not targeted at readership in a particular country, and provides an international perspective. However, not all sections of the book are relevant to all countries. For example, the discussion of choice of technology is unlikely to be relevant to the high income countries of Western Europe or North America; the material on winter maintenance is not relevant to countries in Africa and South East Asia. Nevertheless, most of the book deals with fundamental principles, and should be widely applicable.

Acknowledgements

The book is a textbook and not an encyclopedia. It has attempted to focus on those areas where new or innovative concepts are known to exist, and for which there is a lack of existing textbook material. Several aspects of road maintenance management are not covered, including bridges and structures, footways, footpaths and cycle tracks, street lighting, traffic control and signals, and road safety. Environmental issues are described only briefly. The areas of work included are those where the authors have most experience, based on their work in road administrations, consultancy and research.

However, the authors have included other material where this has been seen as innovative and complementary to the basic text. This work has been referenced in the text, as appropriate, but the authors wish to acknowledge the use made of material from a number of sources. This includes material emanating from the World Bank: in Chapter 3 on pricing and cost recovery that is based on work by Ian Heggie; Bill Paterson's work on information quality and data design included in Chapter 5; and Jean-Marie Lantran's work on equipment management and procurement included in Chapters 8 and 9. We are very grateful for permission from the World Bank to quote from this work. We have also drawn heavily on work on cost estimation carried out by the University of Manchester Institute of Science and Technology on behalf of the then United Kingdom Overseas Development Administration (now Department for International Development (DFID)), and included in Chapter 4. Again, we are very grateful to DFID for permission to quote from this work.

The material in Chapters 6 and 7 is based on a review undertaken by one of the authors for the Transport Research Laboratory. The material on data management in Chapter 5 is based on guidelines produced for the European Commission by May Associates to which one of the authors made a substantial contribution.

Every effort has been made to trace all copyright holders, but if anyone has been overlooked the publishers will be pleased to make the necessary arrangements at the first opportunity.

In addition, we would like to acknowledge the contribution of the many individuals who over the years have knowingly or otherwise contributed to the development of the ideas in this book. These include: Mathew Betz, Barry Blunt, (the late) David Brooks, Neville Bulman, Jeff Falconer, Clell Harral, Dave Hattrell, Ian Heggie, Tom Jones, Henry Kerali, Nick Lamb, Keith Madelin, Peter May, (the late) Ray Millard, Kevin O'Sullivan, Neville Parker, Bill Paterson, Peter Roberts, Bent Thagesen, Sydney Thriscutt and Keith Youngman. We apologise to anyone inadvertently omitted.

The original idea for the book came from the Jubilee Professorship at the Royal Institute of Technology in Stockholm, a post funded by the Swedish National Road Administration (SNRA). We are particularly grateful for the contribution of funding from SNRA without which this book would not have been started.

RR, UD and MSS   *Birmingham UK*

1 Management Issues

- Purpose of management
  - roads as an asset
  - activities
  - impacts
- Construction and maintenance
  - projects and processes
  - purpose of maintenance
- The management process
  - management functions
  - project management
  - the management cycle
- Quality management
  - the concept of customers
  - quality management systems
  - risk management
- Management information

1.1 Purpose of management

1.1.1 Roads as an asset

An efficient road transport system is seen by most countries as an essential pre-condition for general economic development, and considerable resources are devoted to road construction and improvements. The resultant road networks usually have an asset value that represents a significant proportion of national wealth, and the road sub-sector should make an important contribution to gross national product (GNP). It is, therefore, important and appropriate that this asset is managed in a business-like manner, and the attention of an increasing number of road professionals is now being directed to this activity.

Road management, like any management activity involves the following tasks (Adair, 1983):

- Defining activities
- Planning
- Allocating resources
- Organising and motivating personnel
- Controlling work
- Monitoring and evaluating performance
- Feeding back results to seek improvements.

In particular, road management has the purpose of maintaining and improving the existing road network to enable its continued use by traffic in an efficient and safe manner. In addition, appropriate management must also take into account issues of effectiveness and concern for the environment. Road management can be seen as a process that is attempting to optimise the overall performance of the road network over time. In other words, the process may be seen to comprise a number of *activities* (or measures) that will have *impacts* (or effects) on the road network.

1.1.2 Activities

Activities include works to undertake road maintenance, new construction, improvements and the like. A convenient classification of the different types of works is given in Table 1.1. This considers activities in terms of their frequency of application, the budget head used to fund them and their impact on the road infrastructure.

Table 1.1 *Works activities*

Works category	Works type	Description	Examples of works activities
Routine • Works that may need to be undertaken each year	Cyclic	Scheduled works whose needs are dependent on environmental effects rather than traffic	Vegetation control Clearing side drains Clearing culverts
• Normally recurrent budget	Reactive*	Works responding to minor defects caused by a combination of traffic and environmental effects	Crack sealing Patching Edge repair
Periodic • Planned to be undertaken at intervals of several years	Preventive	Addition of a thin film of surfacing to improve surface integrity and waterproofing that does not increase the strength of the pavement	Fog seal/surface rejuvenation Slurry seal
• Typically recurrent or capital budget	Resurfacing	Addition of a thin surfacing to improve surface integrity and waterproofing, or to improve skid resistance, that does not increase the strength of the pavement	Single surface dressing Porous asphalt Thin overlay
	Overlay	Addition of a thick layer to improve structural integrity and to increase the strength of the pavement	Dense-graded asphalt overlay Bonded concrete overlay Regravelling unpaved roads

	Pavement reconstruction	Removal of part or all of the existing pavement and the addition of layers to restore or improve structural integrity and to increase the strength of the pavement	Inlay Mill and replace Full pavement reconstruction (asphalt or concrete)
Special works • Frequency cannot be estimated with certainty in advance	Emergency	Works undertaken to clear a road that has been cut or blocked	Traffic accident removal Clearing debris Repairing washout
• Typically special or contingency funds, but sometimes recurrent budget	Winter	Works undertaken to prevent the formation of ice or to remove snow from the pavement	Salting/gritting Snow removal
Development • Planned at discrete points in time • Normally capital budget	Widening	Works that retain the existing pavement, but increase width throughout the length of the section	Shoulder improvements Partial widening Lane addition
	Realignment	Works that change the road geometry for part of a section, but that retain some of the existing pavement structure	Junction improvements Local geometric improvements
	New section	Works to create a new pavement in a new location	Dualling Construction

* 'Reactive' works have, in the past, sometimes been termed 'recurrent', but this terminology can cause confusion with the *recurrent budget*; consequently, the term *reactive* is now preferred.

1.1.3 Impacts

Road transport is a key component of the economic and social development process, often absorbing a high proportion of national budgets. There is a strong correlation between the kilometres travelled and the GNP (Madelin, 1996). It aids development by facilitating trade both nationally and internationally, and by improving access of people to jobs, education, health care, education and other services. Effective and efficient road transport lowers input prices and, hence, production costs, and can lead to greater economic well-being.

Improvements in the quality of service provision increase personal mobility and facilitate economic growth. These, in turn, contribute towards social development and, particularly in the poorest countries of the world, assist in reducing poverty. Thus, both the quantity and quality of road transport infrastructure have impacts on all aspects of life. Higher mobility is not only a result of development, but also enhances development; it is evident that high income countries are also countries with high general mobility. Small improvements in the costs of maintaining the physical infrastructure, and in the costs of road transport provision and operation, can result in large economic benefits. Typically, $1 of expenditure on maintenance results in $3 of savings to road users (Heggie, 1995).

The benefits derived from road transport are not, however, without their problems. There may sometimes be substantial minorities of people who suffer as a consequence. For example, many people are killed and injured in road accidents, or suffer through pollution and environmental damage. The challenge of road management in a rapidly changing world is not just to minimise adverse impacts on disadvantaged groups or on the environment, but to identify how best to use roads as a mechanism to support economic and social development, and to effect environmental improvement. As a minimum, the management of road infrastructure must facilitate growth and channel the derived benefits to those in most need.

Thus, it can be seen that the impacts of road activities include:

- Level of service or road condition
- National development and socio-economic impacts
- Road user costs
- Accident levels and costs
- Environmental degradation
- Road administration costs.

Road management can be considered to be a process that impacts in these areas, over time, by expending resources on activities.

1.2 Construction and maintenance

1.2.1 Projects and processes

The preoccupation of road professionals in the past has been mostly with road construction. However, as road networks around the world are substantially completed, and as existing networks age, the emphasis of work is now changing away from construction and towards maintenance. Compared with construction, the problem of managing road maintenance has proved to be a particularly difficult issue for many countries (World Bank, 1988; Schliesser and Bull, 1993). Road networks are, by their nature, spread over a wide geographic area, and their condition is changing continuously as a result of the effects of traffic, climate and environmental conditions. The effect on this of maintenance interventions is complex, and is often difficult to predict, particularly over the longer term. It can be seen that maintenance is, essentially, a management problem concerned with issues of:

- Delivering a defined quality of service
- Resources of people, materials and equipment
- Activities and procedures
- Location on the road network
- Timing of interventions.

In addition, road professionals have in the past been concerned with the management of construction projects. But road maintenance is not a *project*: projects have a clearly defined beginning and end, and require the consumption of resources to move from the start to the finish. Although maintenance consumes resources, it is an ongoing activity with no start or end. As such, it is a *process* rather than a project. Whereas traditional engineering management uses the techniques of *project management*, it can be seen that road management needs to draw on the wider resources of business management if it is to be undertaken effectively. Some of the differences between road construction and maintenance are summarised in Table 1.2.

Table 1.2 *Some differences between construction and maintenance*

	Construction	Maintenance
Nature of activity	Project	Process
Duration	Tend to be short term	On-going/long term
Location	Relatively constrained	Normally widespread
Cost per kilometre	Relatively high	Relatively low
Principal skills required	Engineering, project management	Engineering, business management

However, it should be noted that the *process* of road management will contain activities that can often be undertaken on a *project* basis. For example, works to provide an overlay or to repair a bridge can be considered as discrete projects, as can the clearing of a particular length of ditch. There is a considerable amount of literature published on road engineering (see, for example, Oglesby and Hicks, 1982; Gichaga and Parker, 1988; Salter, 1993; Thagesen, 1996) but, although some of these discuss the management of road maintenance, there is relatively little available that focuses exclusively on this subject.

1.2.2 Purpose of maintenance

Maintenance reduces the rate of pavement deterioration, it lowers the cost of operating vehicles on the road by improving the running surface, and it keeps the road open on a continuous basis (World Bank, 1988). It also includes the process of enhancing the environment of the road itself, including the immediate surroundings. Maintenance should also be carried out to improve safety but, paradoxically, this is sometimes

6 *Road Maintenance Management*

problematic as it can lead to increased speeds which, in turn, result in increased numbers and severity of accidents.

Within this broad purpose, maintenance management can be assumed to have more detailed aims (Local Authority Associations, 1989). These include:

- The use of a systematic approach to decision-making within a consistent and defined framework.
- To assess budget needs and resource requirements.
- To adopt consistent standards for maintenance and for the design of associated works.
- To allocate resources effectively.
- To review policies, standards and the effectiveness of programmes on a regular basis.

An example of the relative costs over the lifetime of a typical road in Sweden is shown in Figure 1.1. From this it will be seen that the cost of maintenance activities is very small when compared with other costs. Nevertheless, the *impacts* of maintenance on these other costs can be very significant. Similarly, the benefits in other areas can be substantial as a result of relatively small expenditures on road maintenance. Maintenance costs per vehicle-kilometre, also from Sweden, are shown in Figure 1.2. An example of the relative costs of the different types of works is shown in Table 1.3. The difference in the works costs of paved and gravel roads will be noted.

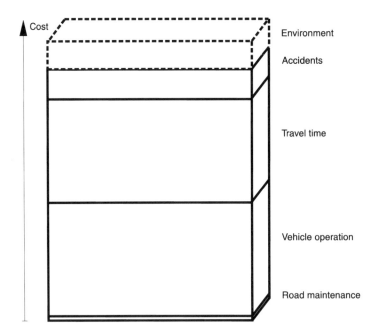

Figure 1.1 *Relative costs on an ordinary road in Sweden* (*Source*: Swedish National Roads Administration)

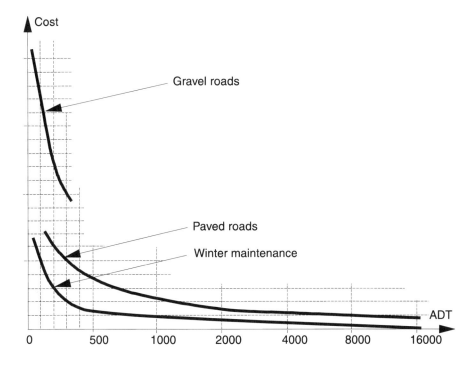

Figure 1.2 *Road maintenance costs per vehicle-kilometre (Source: Swedish National Roads Administration)*

Table 1.3 *Example of relative costs of road management activities*

	Cost ($US/km)
Paved roads	
Routine maintenance	1700 + 0.5T
Periodic reseal	22 400
40 mm overlay	56 000
80 mm overlay	90 000
Pavement reconstruction	280 000
Gravel roads	
Grading	$US80/km/grading
Regravelling	$US7/m^3

Notes: The figures quoted are an average over a number of World Bank projects.
T = average annual daily traffic (vehicles/day).

Source: Heggie (1995).

Maintenance covers a wide range of activities, many of which lack the 'glamour' associated with new works. As such, maintenance is not spectacular and, sometimes, its results do not have immediate impact. The long-term effects of maintenance are, however, significant. A key challenge for the road manager is to find ways in which to describe the problems and impacts of road maintenance that can be understood by politicians and the general public. It is very much more difficult to describe and define road conditions when maintenance intervention is necessary than to describe conditions resulting from construction: describing and measuring something that is 'bad' is much more difficult than for something that is 'good'. Doing this successfully is vital if adequate funding and support are to be obtained. This issue is considered again under 'benefits' in Chapter 4.

1.2.3 Reducing deterioration

Even with adequate maintenance, pavements will deteriorate over time. The rate of deterioration will depend on a number of factors including the traffic loading, the pavement strength, the climate and the environment. Eventually, the end of a pavement's design life will be reached and there is a need for pavement reconstruction or upgrading. These are normally relatively expensive activities and should, therefore, be postponed for as long as possible by carrying out effective and timely maintenance.

If the required cyclic and reactive maintenance are not carried out, drainage will become ineffective and surface defects will worsen, both of which result in water penetrating the structure of the pavement. For paved roads, the resulting distress requires that a higher level of maintenance is needed prematurely. Failure to carry out resurfacing maintenance at the appropriate time soon leads to the need to carry out strengthening overlay works, which is at least twice as expensive as resealing. If this overlay is not carried out soon enough, major deterioration sets in and pavement reconstruction will be required, which is at least three times more costly than an overlay. It will be seen that deferring works results in a rapid escalation of costs to the road administration.

The effect of axle loading and, in particular of overloaded vehicles, on the requirement for road maintenance is considerable. For example, a 10 tonne axle causes approximately 2.5 times as much deterioration to a pavement as an axle weighing 8 tonnes (Liddle, 1963). It is clearly necessary, for road maintenance purposes, to know the value of the actual axle loading, since minor underestimates can shorten considerably the expected, and hence designed, life of a pavement.

However, enforcement of axle loading legislation is often difficult because the incentive system is biased: individual road users benefit from overloading at the expense of road users as a whole. Thus there is no incentive for individual users to comply other than the threat of prosecution. From a road management point of view, there is considerable advantage in having appropriate axle load legislation which is enforced effectively.

1.2.4 Lowering vehicle operating costs

Cost savings obtained by deferring the need for reconstruction, quoted above, exclude any benefits to vehicle operators who thereby avoid the high costs of operating on badly deteriorated pavements. The relative proportions of road administration costs and vehicle operating costs in the total lifetime transport cost associated with a road vary depending on the traffic level, as shown in Figure 1.3. This figure is based on research carried out by the World Bank and relates to roads where optimal maintenance is undertaken (Schliesser and Bull, 1993). This shows that the relative proportion of vehicle operating costs rises from about 40 per cent at 50 vehicles/day to over 90 per cent at 6000 vehicles/day.

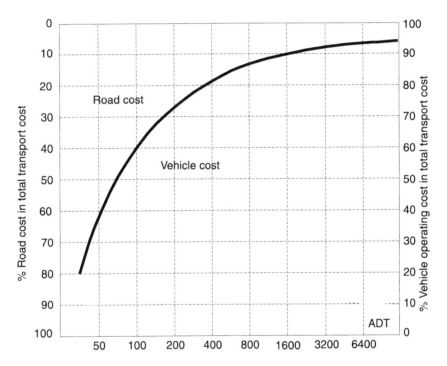

Figure 1.3 *Relative proportions of road and vehicle costs in the total transport cost* (*Source*: Schliesser and Bull, 1993)

A further example, shown in Figure 1.4, illustrates the effect of neglecting road maintenance. The figure shows the relative discounted life cycle costs of construction, maintenance and vehicle operation under different maintenance spending scenarios. For a traffic level of about 1000 vehicles/day, a road in good condition will require about 2 per cent of the total discounted costs to be spent on maintenance. However, if maintenance funds are reduced, the pavement will start to crack and potholes will gradually appear. With this level of deterioration, vehicle operating costs are likely to increase by about 15 per cent. If there is complete neglect of maintenance, a paved road will eventually start to disintegrate, and annual vehicle operating costs will increase by about 50 per cent.

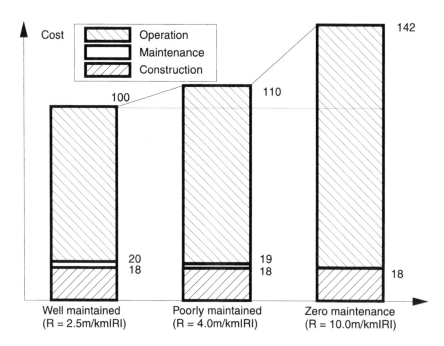

Figure 1.4 *Change in discounted life cycle costs on a paved road for different levels of maintenance* (*Source*: Robinson and Roberts, 1982)

1.2.5 Keeping the road open

The third reason for carrying out maintenance is to keep the road open continuously. Roads serve centres of population and industry and, if roads are closed, for whatever reason, then there are potentially serious social and economic consequences.

In the case of winter maintenance, decisions need to be taken about where and when to remove snow. It is impracticable to provide 'summer-like' conditions at all times during winter months, and compromises may have to be made. This may involve decisions not to clear snow from most minor roads, or to keep only a limited number of lanes open on multi-lane roads.

1.2.6 Safety

Accidents have proved to be an inevitable result of road transport, and deaths and injuries are very tangible impacts of roads on the community. The factors contributing to safety are engineering, education and enforcement. In many countries, the climate also has a significant impact. In this context, education aims at changing behaviour through publicity and raised awareness. While education and enforcement fall outside the scope of this book, it must be appreciated that all four factors interact, and that different combinations of factors are likely to have different impacts. However, in extreme cold climates, the special requirements associated with them will dominate.

A co-ordinated approach to safety should, therefore, be adopted at national, regional and local levels.

Road maintenance works can often provide an opportunity for making improvements to road safety by contributing to the *engineering* factors in the areas of:

- Pavement and footway surfaces
- Carriageway markings and delineation
- Signs, street lights and road furniture.

As a result, safety considerations should influence the determination of need and the approach to maintenance in all of the above. Good practical advice on this is contained in documents published by the Institution of Highways and Transportation (IHT, 1990) and the Transport Research Laboratory (TRRL Overseas Unit, 1991). These address issues of 'black spot' analysis, remedial works design and layout, and traffic management measures.

1.2.7 Environmental issues

The condition of roads also affects the environment (World Bank, 1994). This is important in all cases, and there is a growing public expectation for their surroundings to be managed properly. As noted above in sub-section 1.2.4, roads in poor condition also lead to wasted non-renewable resources and contribute to air pollution from vehicles that are not operating efficiently. A further example is the need to consider the noise characteristics of different pavement treatments at the treatment selection stage, particularly in sensitive areas, such as those adjacent to hospitals or schools. Similarly, street lighting plays a vital part in crime prevention and the safety of vulnerable groups.

Particular problems also arise with chemical pollution in surface water running from roads. This may contain tyre detritus, diesel spillage, salts and other undesirable substances. Road drainage systems need to be designed to cope with such effluent. Other design, construction and maintenance problems include the need to conserve primary materials resources.

1.3 The management process

1.3.1 Management functions

It is convenient to consider the road management process in terms of the following primary functions:

- Planning
- Programming
- Preparation
- Operations.

These functions can be described as follows.

Planning

This involves an analysis of the road system as a whole, typically requiring the preparation of long-term, or strategic, planning estimates of expenditure for road development and conservation under various budgetary and economic scenarios. Predictions may be made of expenditure under selected budget heads, and forecasts of road conditions, in terms of key indicators, under a variety of funding levels.

The physical road system is likely to be characterised at the planning stage by lengths of road, or percentages of the network, in various categories defined by parameters such as road class or hierarchy, traffic flow or congestion, pavement type and physical condition.

The results of the planning exercise are of most interest to the senior policy makers in the road sub-sector, both political and professional. Work will often be undertaken by a planning or economics unit.

Programming

This involves the development, under budget constraints, of multi-year works and expenditure programmes in which those sections of the network likely to require treatment, and new construction possibilities, are identified and selected. It is a tactical management exercise. Ideally, a cost–benefit analysis should be undertaken to determine the economic feasibility of each set of works.

The physical road network is likely to be considered at the programming stage on a link-by-link basis, with each link characterised by pavement sections and geometric segments, each of which is defined in terms of physical attributes. The programming activity produces estimates of expenditure, under different budget heads, for different treatment types and for different years for each road section. Budgets are typically constrained, and a key aspect of programming is to prioritise works to find the best value for money in the case of a constrained budget. Typical applications are the development of a budget for an annual or rolling three-year works programme for a network, or perhaps a sub-network of national roads.

Programming activities are normally undertaken by managerial-level professionals who may be budget holders in the organisation, perhaps in a planning or a maintenance department.

Preparation

This is the stage where road schemes and projects are packaged for implementation. At this stage, designs are refined and prepared in more detail; bills of quantities and detailed costings are made; together with work instructions and contracts. Detailed specifications and costings are likely to be drawn up, and any cost–benefit analysis may be revisited to confirm the feasibility of the final project or scheme.

Works on adjacent road sections may be combined into packages of a size that is cost-effective for works execution. Typical preparation activities are: the detailed design of an overlay project or scheme; the preparation of tender documents and the letting of contract; or other such items. For these activities, budgets will normally already have been approved.

Preparation activities are normally undertaken by engineers and technicians in a technical department of an organisation, and by contracts and procurement staff.

Operations

These activities cover the on-going works activities of an organisation. Decisions about the management of operations are made typically on a daily or weekly basis, including the scheduling of work to be carried out, monitoring in terms of labour, equipment and materials, the recording and evaluation of work completed, and the use of this information for monitoring and control.

Activities are normally focused on individual sections or sub-sections of road, with measurements often being made at a relatively detailed level.

Operations are normally managed by works supervisors, technicians, or clerks of works.

As the management process moves from *planning* through to *operations*, it will be seen that the following changes occur (developed from Paterson and Robinson, 1991):

- The view of the road sections changes from the overall network to only those sub-sections where works are being carried out.
- The time horizon being considered changes from multi-year to budget year, and then to the current week or day.
- The level, experience and grade of staff concerned change from senior management to technicians and supervisors.
- The definition of works in Table 1.1 changes from 'categories', through 'type' to 'activities'; the term 'task' would normally be used to define in detail works undertaken at the operational level.
- Management data change from being summary or sampled, to detailed with full coverage of the particular projects or schemes; data accuracy increases.

These changes are summarised in Table 1.4.

1.3.2 Project management

Note that the terms planning, programming, preparation and operations refer to the *process* of management. However, works themselves can be considered as discrete *projects*, whether they are undertaken by the in-house organisation or under contract. In the context of projects, different terminology for the management functions is normally used (TRRL Overseas Unit, 1988a):

1. *Identification*
   Identification of a project.
2. *Feasibility*
   Appraisal of the identified project, including basic requirements, alternative projects, and recommendation of preferred project; normally requires a cost–benefit analysis.

3. *Design and commitment*
   Definition of the preferred project, including basic design data, technical specifications, construction appraisal, contract strategies, and estimate of final cash cost; consideration of submission for funding.
4. *Implementation, operation and evaluation*
   Implementation of approved project, including
   - detailed design
   - issue of tender enquiries
   - assessment of tenders
   - placement of contracts
   - construction
   - completion
   - commissioning
   - operation of new asset
   - evaluation of project.

Table 1.4 *Change in management processes*

Management function	Spatial coverage	Time horizon	Staff concerned	Definition of works[1]	Data
Planning	Network-wide	Long term (strategic)	Senior management and policy level	Works category	Coarse/ summary
Programming	Network to sub-network	Medium term (tactical)	Middle-level professional	Works type	
Preparation	Section or project (or scheme)	Budget year	Engineer/ technician	Activity	
Operations	Sub-section	Intermediate/ very short term	Technicians/ works supervisor	Task	Fine/detailed

(1) see Table 1.1.

It will be seen that there is a one-to-one relationship between the management functions and the steps used to execute a project, as shown in Table 1.5. Other chapters of the book show how various aspects of management fit within this same framework. This includes aspects such as policy, environmental assessment, cost estimation, data detail and ease of contracting out maintenance activities. These relationships are summarised in Appendix A.

1.3.3 The management cycle

Traditionally, in many road organisations, budgets and programmes for road works have been prepared on a historical basis, in which each year's budget is based upon that for the year before, with an adjustment for inflation. Under such a regime, there is no way of telling whether funding levels, or the detailed allocation, are either adequate or fair. Clearly, there is a requirement for an objective *needs-based* approach, using knowledge of the content, structure and condition of the roads being managed. It will be seen that the primary functions of *planning, programming, preparation* and *operations* provide a suitable framework within which a needs-based approach can operate.

Table 1.5 *Relationship between management functions and project steps*

Management functions	Project steps
Planning	Identification
Programming	Feasibility
Preparation	Design and commitment
Operations	Implementation, operation and evaluation

Within each of the four primary management functions, a common system can be used. An appropriate approach is to use the concept known as the *management cycle*, which is illustrated in Figure 1.5. The cycle provides a series of well-defined steps that take the management process through the decision-making activities. The process typically completes the cycle once in each period covered by the management function.

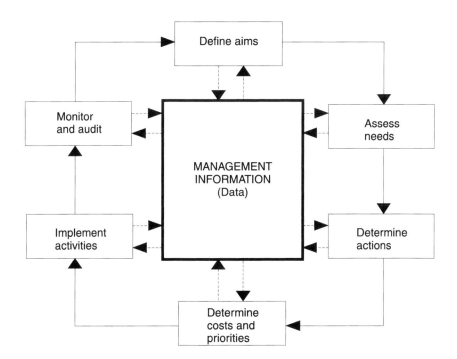

Figure 1.5 *The road management cycle*

As the management of any activity is only possible with appropriate and up-to-date information, road management information should be considered to sit at the heart of the management cycle. The decisions that are made in each step of the cycle use this information. Hence it is the combination of decisions and information that enables a management strategy to be put in place.

The following paragraphs introduce each box from Figure 1.5 in turn. Note that the detailed content of each step will depend on which of the four management functions the management cycle is being used to address.

Define aims

The overall policy of the road administration is clearly important in road management. The policy framework of the administration defines its aims for road management. Defining the goal being sought is an essential first step in the management cycle for each of the management functions, since all activities must be undertaken with a view to meeting defined aims.

Assess needs

The key step in assessing needs is the collection of data. These data provide information on the extent of the gap between the present level of service or standard and that set out in the aim being sought, as defined above. This, in turn, permits identification of the activities that are required.

Determine actions

There will often be choices to be made about the activities required to meet the aims identified by the above. Alternatives must be considered to determine the most appropriate options.

Determine costs and priorities

Actions identified above must be costed to identify the resource requirements. Resource needs identified will normally be greater than those available to carry out the required activities. A rational system of setting priorities is therefore required to allocate available resources in a systematic and equitable way, such that the best value for money is obtained.

Implement activities

This involves all activities undertaken during execution of the management functions. Implementation must ensure that activities are carried out to pre-defined standards in order that the administration's aims are met, and supervision is an important component of this.

Monitoring and audit

A review process must form an integral part of needs-based road management. This process should include the following two activities:

- *Monitoring*
  This has the principal function of providing feed-back to the management process, so that when the next cycle of management takes place, it can learn from past experience. For example, aims can be redefined to reflect the actual achievements; unit rates can be revised to reflect those actually obtained in the field; or, indeed, technical methods may be improved on the basis of the monitoring.
- *Audit*
  This includes both technical and financial audit, and provides a physical check, usually on a sample basis, that work has been carried out, where specified, to pre-defined standards or procedures, and that costs and other resources have been accounted for properly.

As suggested by the word 'cycle', the *management cycle* is a repetitive process which, for most management functions, is undertaken once for each period of the management function. Thus, for example, the management cycle for programming would normally be carried out once per year; whereas that for operations would typically take place once every one or two weeks.

1.3.4   Cycles within the management functions

Considering the road management process as a whole can, therefore, be considered as a cycle of activities that are undertaken within each of the management functions of *planning, programming, preparations* and *operations*. This is conceptualised in Figure 1.6, and provides the framework within which the road management process operates. Examples of management cycles for each of the four management functions are given in Table 1.6. That for the programming function is illustrated in Figure 1.7.

1.4   Quality management

1.4.1   The concept of customers

As noted previously, the focus of most road organisations has been traditionally on design and construction of pavements and structures. However, the only reason for providing roads is to enable them to be used by people. It is, therefore, the needs of the road users that should dictate the policy and objectives for the road organisation. It is convenient, for these purposes, to think of the road organisation in terms of a commercial business which should be trying to address *customer* needs, where the *customers* are the road users. Such considerations are entirely consistent with the precepts of quality assurance and quality management as enshrined in *ISO 9000* (ISO, 1987).

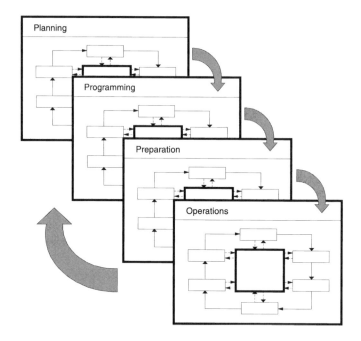

Figure 1.6 *The management functions and cycles*

Table 1.6 *Examples of management cycles for different functions*

Steps in the management cycle	Management functions			
	Planning	Programming	Preparation	Operations
Define aims	(a) Determine standards that minimise total costs (b) Determine budget required to support given standards	Determine works programme that can be carried out with next year's budget	(a) Design of work (b) Issue of contract or work instruction	Undertake works activity
Assess needs	Assessed using surveys of aggregate network condition for periodic and some reactive works, and using historical data for cyclic and special works	Assessed by comparing condition measurements with standards for periodic and some reactive works, and historical data for cyclic and special works	(a) Assessed by undertaking detailed surveys to assess condition and compared with design standards for periodic and development works (other works not	Extent and quantity of work are assessed: • From detailed inspections for reactive and periodic works • From the standard for cyclic works

			normally designed) (b) Appropriate form of contract or work instruction selected	(which depend on the nature of the works for special works)
Determine actions	(a) Treatments determined by applying a range of standards to give budget requirements spanning the required range (b) Treatments determined by applying fixed standards	Works options that are available to restore conditions to standards are determined	(a) Design options determined (b) Options for specifications determined	Appropriate performance standard is selected for the activity
Determine costs and priorities	a) Application of cost rates to give budget requirements, with treatments prioritised to meet budget constraints (b) Application of cost rates to determine budget requirements, with no prioritisation	Cost rates are applied and options prioritised to produce a programme within the budget	a) Cost rates are applied and priorities are possibly reconfirmed b) Bill of quantities prepared	Apply targets, and labour, equipment and material resource requirements from standard
Implement actions	(a) Publish standards (b) Publish forecast budget needs	Submit works programme	(a) Undertake design, produce drawings, etc. (b) Prepare and let contracts or issue work instructions	Undertake and supervise work
Monitor and audit	• Review forecasts prior to start of next planning cycle • Review planning procedures	• Review programme produced prior to start of next cycle • Review programming procedures	• Review or check design, or contract or work instruction • Review design procedures	• Review achievement against target • Review procedures for managing works activities
Length of cycle	Typically 3–5 years	Typically one year	Typically less than year	Typically days or weeks

Quality can be defined as conformance with customer requirements. The aim of quality management is to improve business performance by understanding customer requirements, and then controlling the various activities to ensure that those requirements are met first time, every time. Note that quality assurance differs from quality

20   *Road Maintenance Management*

control in that it includes every management activity (the causes), and not just the inspection of the end products (the effects). In order to deliver quality through their road management services, road administrations need to reflect customer requirements in briefs, procedures and plans so that they can be understood and agreed. Managers are then responsible for helping their teams to meet these and any other requirements.

1.4.2   Quality management systems

To avoid confusion, it is convenient to define terms that are commonly used in connection with the quality management of processes (Munro-Faure *et al.*, 1993).

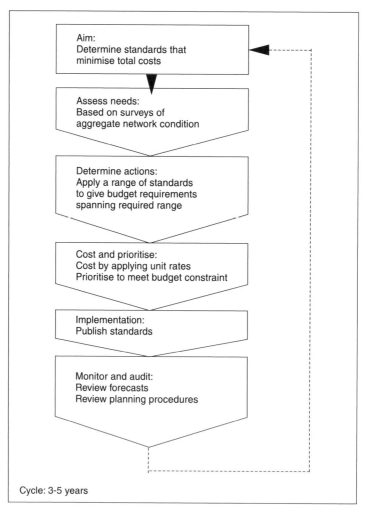

Figure 1.7   *Example of the management cycle for programming*

Quality

The totality of features and characteristics of a 'product' or service that bear on its ability to satisfy *stated* or *implied* needs.

Quality assurance (QA)

All those planned and systematic actions necessary to provide confidence that a particular product will satisfy given end-product requirements for quality.

Quality control

The operational techniques and activities that are used to check that requirements for quality have been met.

Total quality management (TQM)

The management approach of an organisation, centred on quality, based on the participation of all of its members, and aiming to achieve long-term success through customer satisfaction, and benefits to all members of the organisation and to society more generally.

A quality assured management system contains the components shown in Box 1.1. An example of the use of this approach in the road management area is given by Zimmermann and Darter (1994). Quality assurance provides a framework for the management of all activities within a road organisation, and for introducing sustained improvements to operations.

Box 1.1 **Components of a quality assured management system**

Management responsibilities
Clearly defined roles and responsibilities for all those involved

Quality procedures
Documented and auditable procedures for all activities

Operations control
Control of all activities, documents and the management of change

Inspection and testing
Formal checking and approval of all activities, documents and equipment

Records
Procedures for the maintenance of records

Audit
Regular audit and corrective action

Adapted from: Roberts and Pritchard (1991).

1.4.3 Risk management

All aspects of road management involve uncertainty and risk; all are set against a background of economic, social, political and engineering uncertainty. Effective management involves the collection of significant amounts of data and the forecasting of trends into the future. All data collected in the field are subject to errors, and some may be extremely inaccurate. By the time these data have been extrapolated to make future projections, any errors present may have been magnified significantly. When this is coupled with uncertainties that exist in the projection process itself, it will be clear that such decisions may be subject to substantial errors.

As uncertainties always exist, it is necessary to take steps to minimise them. It is also necessary to determine the effect of uncertainty on the robustness of the conclusions reached as this will affect the final recommendations. The method of dealing with this will depend on the time frame involved. If short-term forecasts are being used, for example, for budgeting, then the use of *risk analysis* may well be appropriate. This involves formal probability analysis of the likely range of outcomes. *Sensitivity analysis* is useful for identifying critical variables. If longer range forecasts are being made, such as in the case of network planning, then *scenario analysis* may be more appropriate. Another approach to risk management is through *value engineering*.

Risk analysis

All variables are assigned probabilities for the likelihood of different values occurring. Probability analysis is carried out to determine the resulting outcomes, which are also produced as a range of values indicating probabilities of different outcomes arising. In this way, the magnitude of the major uncertainties is highlighted. Many formal methods are available for carrying out risk analysis, and examples of its application to road projects are given by Pouliquen (1970) and Ellis (1975).

Sensitivity analysis

This process works by evaluating the outcome when each of the more important variables is changed in turn by a fixed amount, but keeping the values of the remaining variables fixed. By examining the relative effects on the outcome, the most sensitive variables can be identified.

Scenario analysis

This differs from sensitivity analysis because several variables are changed at the same time to give a scenario for which the outcome can be determined. By investigating several scenarios, the range of likely outcomes can be determined. Scenarios are selected to reflect the range of future possibilities that might reasonably be expected to occur. The choice of scenarios is made such that they span the plausible range of possible outcomes, rather than to provide a forecast of what will actually occur. The relevant principles of scenario analysis, as relevant to the sector, are described by Allport *et al.* (1986).

Value engineering

Those variables having the greatest impact on the final costs are identified, and the likely variability in these is estimated. Effort then focuses on reducing the variability of the most significant variables, which may or may not be worthwhile depending on the cost of the investigation and the expected reduction in risk. Value engineering can also identify where risk might be reduced by adopting a more flexible approach to decision making (AASHTO, 1987).

Risk management provides a basis for judging the relative merits of alternative decisions, but, in itself, does nothing to diminish the risk. The results of the risk analysis need to be taken into account in the engineering decision-making process. For decisions that are as complex as those involved in road management, risk management is time-consuming to carry out. As such, it should be reserved for a very few parameters in highly critical cases. Some risks identified can be reduced by carrying out further field investigations and, possibly, redesign. However, for a small amount of effort, rough-and-ready forms of risk analysis are likely to improve the quality of decision making considerably (Hambly and Hambly, 1994).

1.5 Management information

It has already been noted that effective management requires appropriate and up-to-date information to support management decisions. The need to assess physical condition, safety, level of service, and efficiency of operation of road systems is widely recognised.

In addition to knowing the characteristics of the existing system, it is becoming increasingly important to be able to predict the effects that proposed policies are likely to have in the future. Such predictive capabilities enable the decision maker to test alternative courses of action to determine which policies and strategies will be the most effective in accomplishing the desired goals with the resources available.

Data are, therefore, needed to provide the basis for management decisions on such aspects as:

- Assessing current levels of road and bridge condition
- Determining appropriate levels of investment
- Prioritising capital improvements and investments in maintenance
- Simulating the effects on any improvements on the future condition and performance of the road system
- Estimating the cost of improvements
- Preparing designs that reflect the impact of time on the decisions, by being maintenance-friendly
- Identify the life cycle cost impact of preventive works in delaying the need for major works
- Controlling on-going expenditures.

Management information of this nature provides the quantitative basis of dialogue between technical departments and ministries of finance during the fund allocation process. It can also provide a quantitative basis of a dialogue with elected representatives, and for monitoring departmental performance and the meeting of policy objectives.

The need to process data and provide meaningful information through management reports for the direction and control of road operations can be greatly facilitated by using computers. The systems need to be structured to support the primary management functions of: *planning, programming, preparation* and *operations*. It is the role of a road management system to provide both a knowledge base and decision support to assist in the generation and evaluation of management strategies for judging the adequacy of overall investment; they are used at the tactical level for determining budget needs and priorities; and they are used at the detailed level for project preparation and the management and control of operations and budgets.

Road management systems are described more generally in Chapter 5, and some models associated with their use are described in Chapters 6 and 7.

2 Institutional Issues

- Management in context
  - external and internal factors
  - economic factors
  - roles and aims
  - stakeholders
- Legal ownership
- The policy framework
  - management and policy
  - vested interests
  - mission statement, objectives, standards and intervention levels
  - functional hierarchy
  - performance indicators
- Organisations
  - organisational culture
  - size and degree of decentralisation
  - organisational status
- Effectiveness and efficiency
  - specificity and competition
  - functional separation
  - commercialisation
- Institutional development
  - hierarchy of management issues
  - the challenge of change
- Human resource management
  - human resource needs in a changing environment
  - skills and motivation
  - training
- Environmental issues

2.1 Management in context

2.1.1 External and internal factors

The manner in which road management is undertaken, and the constraints to which it is subjected, are affected by a number of factors. In particular, the management context is influenced by what may be considered as *internal* and *external* factors relative to the organisation responsible for managing the road network. Internal factors are those that are within the control of the road administration. They can be split into those that are *technical* and those that are *institutional*.

Technical factors reflect an organisation's capability to undertake physical or engineering tasks. They can be considered as the 'hard' internal factors, and include such as the following:

- The availability and use of
  - data
  - materials and supplies
  - plant and equipment
- Capability to undertake
  - technical operations
  - monitoring and feed-back
- Access to research and information.

Institutional factors include organisation and managerial arrangements, finance, human resources and the like. They can be considered as the 'soft' internal factors. They include:

- Finance and resource management, including revenues, budgets and expenditures.
- Organisation and management, including the implementation of policy, organisational and administrative structures, planning, programming and project preparation, and the management of on-going operations.
- Human resources, including the size and composition of the workforce, career development and training and, in many organisations, remuneration and incentives.

External factors are those over which the organisation itself has no direct control, but which constrain the way that the organisation can operate. These factors include:

- The physical environment
- Legal and regulatory framework
- Socio-cultural background of the country
- Political situation
- Macro economy and the national resources available
- Overall government employment policies
- Relationships with other institutions.

For example, the influence of the state of a country's economy and its government policies on investment decisions will impact on the way that the road network develops and on the levels of traffic that will use the network. The economic and political basis of road funding will also impact on many aspects of the operation of the road administration. Factors such as these define the context within which management must operate, but these external factors are beyond the scope of this book (for further information on these aspects see Robinson, 1991a; Schliesser and Bull, 1993; Heggie, 1995).

The way that an organisation handles these issues determines to a large extent how effective and efficient its operations will be.

2.1.2 *Economic factors*

Many countries around the world are able to achieve very high economic rates of return by carrying out maintenance and reconstruction activities rather than by undertaking new road construction. This is because these activities involve relatively low expenditures, but can improve road conditions and give benefits to a large number of road users. New construction is a high cost activity which consequently requires a significantly higher level of benefits to justify expenditure compared with maintenance. This difference is emphasised in many countries around the world where road networks are in poor condition (World Bank, 1988). This implies that, in most cases, maintenance should have higher priority for funding than new construction.

A balance must be struck between maintenance expenditures and reductions in road user costs. For example, on lightly trafficked roads, where only small numbers of road users are incurring costs, the required maintenance expenditures will often outweigh the road user cost savings derived from the provision of a higher level of service. In such cases, it may only be economic to maintain to a very low standard or, in some cases, to carry out no maintenance at all: effectively abandoning particular roads. However, even the smallest roads may have a role in feeding traffic into other parts of the network, and there may be a need to maintain uneconomic roads for socio-political reasons, or to maintain 'regional balance'. Governments should support decisions in this area with appropriate funding.

2.1.3 Roles and aims

It is up to each country to determine the roles and aims of its road administration. Typically, roles and aims will be related to the management of the road network in order:

- To support national socio-economic goals
- To provide acceptable road conditions at a level of service to ensure that roads are safe; to enable transport to operate at minimum cost; and to minimise damage to the environment.

All of these should be achieved within a budget level that can be made available to the road administration. Defining basic aims and objectives in this way enables governments to decide how best to organise the management of the road network.

Many existing road administrations are government departments, so it is relevant to consider the respective roles of government and administration with respect to road management. It is reasonable to assume that the role of government is to govern. To do this requires the definition of policies, and the provision of a legislative framework and the budgets necessary to put these policies into practice. The policy-making function of government is clearly a 'political' function, since the words have the same root. In this sense, it is the role of politics and government to reflect wider national, social and economic interests in defining policy within which a road network can be managed.

Many countries have found it beneficial to separate the political role of government from the professional role of administration (Pontin, 1986). With such an approach in the road sub-sector, government has political responsibility for road policy. The road administration has the responsibility for effecting policy by making decisions based on professional judgement. This may be seen in the example of roles and responsibilities shown in Table 2.1. In addition, the example of the Swedish National Road Administration is presented in Box 2.1.

Table 2.1 *Split of responsibilities between a government and a road administration*

Role of government	Role of the road administration
• Framing policy • Providing a legislative and regulatory framework • Long-term planning, including that between transport modes • Allocation of budgets to the sub-sector • General instructions to the road administration to undertake work on the network	• Maintaining, developing and, more generally, managing the road network • Setting detailed rules and standards for the road network • Execution of the road programme to meet the aims of government • Enforcing regulations for such things as traffic safety and axle loading • Monitoring performance of the road network to improve the way that it is managed in the future, and to inform government on changes to policy that may be advisable

Source: Swedish National Road Administration.

Such an administration can be tasked with delivering, on behalf of government, the mission for the road sub-sector by meeting defined objectives and carrying out work to meet agreed standards. The government ministry can be a small organisation in keeping with its defined function. The administration can be outside the day-to-day control of government ministers, but needs to be publicly accountable for its actions. This is normally through a *roads board* that is autonomous and accountable to parliament or the legislature (Heggie, 1995). The relationship between ministries, parliament and the roads boards is defined by legislation. In general, ministers can publish proposals for new policy directives, and the boards are responsible for carrying out these directives. Boards can make proposals for new policies and can put forward budget proposals. To be really effective, such a system needs freedom of information, public accountability of the boards, and some formal oversight arrangements.

Box 2.1   ***Aims of a typical road administration***

- To be the national road management authority, with the responsibility for planning, design, construction and maintenance of the national road network
- To take the overall, national responsibility for road safety
- To take national responsibility in the road transportation sector for an environmentally sustainable transport system
- To take national responsibility for questions concerning all commercial and international road transport
- To be the government's expert authority on all matters concerning public road transport

Source: Swedish National Road Administration.

2.1.4 Stakeholders

It is important for road administrations to appreciate and understand that there are customers for all aspects of its work (Heggie, 1995; Institution of Civil Engineers, 1996b). The main customers are the road users, who include:

- Owners and operators of commercial vehicles and buses
- Representatives of industry, commerce and agriculture, who have a vested interest in an efficient road network to support their business operations
- The travelling public using the road network.

In addition to customers or users, there are other 'stakeholders' in the road network. These include the road administrations themselves, and the road engineering industry. Road administrations are, normally, a public authority and, as such, are funded directly or indirectly by the tax payer. It is important that their services can be related to the needs of the tax payer, as expressed by government, local politicians and others. Contractors and consultants will be stakeholders where they have an interest in participating in the planning, design and execution of works, and in the management of the road network

Recognising that customer, or user, needs should determine aims for the road sub-sector, it is perhaps worth considering that functions of a road administration are likely to relate most nearly with those of a Ministry of Transport. This ministry should normally be concerned with transport users: those of whom are in the road sub-sector will be the customers of the road administration. Sometimes governmental responsibility for the road administration is placed with the Ministry of Works. However, this places the focus of the administration on the physical infrastructure rather than on the road users, and can therefore give inappropriate policy signals.

2.2 Legal ownership

2.2.1 The concept of ownership

Governments normally aim to create an affordable road network and to have as many roads as possible fully maintained to an acceptable standard. This means establishing an institutional structure that ensures that each part of the road network is owned by a competent road administration. The administration which owns the road normally has some incentive to maintain it, upgrade it when needed, and to ensure that funds for maintenance are spent in a cost-effective manner. Roads without owners are usually neglected (Heggie, 1995).

Establishing ownership normally requires a two-pronged approach which ensures that all the important main, urban and district roads are clearly assigned to a legally constituted road administration, and that *de facto* owners are found for as much as possible of the remainder of the road network. In this way, responsibility for road networks may be assigned to central government, local government, or to a private entity. The approach to establishing ownership is given in Box 2.2.

> **Box 2.2** *Establishing the legal status of roads*
>
> Roads fall into two main legal categories. They are either *proclaimed* or *unproclaimed*. Terminology varies from country to country, and other terms like declared/undeclared and adopted/unadopted are also used to describe legal status. When a road is proclaimed, the act of proclamation is published in the *Gazette*, or other official journal used to record acts of government, in a formal notice. This cites the Act under which the road is to be proclaimed, its location, the road administration responsible, and the functions delegated to the administration. In the case of primary, or trunk roads, the Act cited is usually the Roads or Road Traffic Act, or its equivalent. Urban roads may be proclaimed under the Urban Transport Act, or its equivalent, while other roads may be proclaimed under a variety of other Acts, including the Local Government Act, the National Parks Act, or the Private Street Works Act.
>
> Once a road has been proclaimed, the road administration responsible is expected to mark out the road reserve physically, which defines the land holding of the road administration, and to take responsibility for the various functions delegated to it in the *Gazette* (such as with regard to drainage). Roads that are unproclaimed belong simply to the adjoining land-owners, who are solely responsible for maintaining them. However, under certain circumstances, government may channel funds through a designated road administration to meet part of the costs of maintenance. When a private road is built to a certain specific standard, or is improved to that standard, the government will usually proclaim it and assign it to a legally constituted owner.
>
> *Source*: C. J. Lane and I. G. Heggie, The World Bank.

2.2.2 Ownership and management of capital assets

One approach to road management is to value the road network as an asset. A quantitative approach such as this helps to focus on those aspects of management that help to maintain or increase the value of the asset. Operation of the road administration can then more easily be subject to annual financial and technical audit.

There are different views on the subject of which organisation should manage the road network. But there are several reasons that favour a road administration that manages all publicly owned road assets, excluding the city and municipal streets (Talvitie, 1996). Some of these reasons are:

- There exist economies of scale and scope in building and maintaining roads; efficiency could be improved if one administration was responsible for managing the public road network.
- When all assets are valued, any decision to dis-invest by neglecting maintenance on part of the network is transparent, and becomes a conscious choice.
- Because government expects interest on the capital it invests, the provision and maintenance of roads becomes a socio-economic choice.
- Inclusion of the lower level 'social' road network within the responsibility of the organisation prevents the division of roads, and of society, into two parts – one economic and the other social; the road administration will need to develop fair methods to appraise its investment and maintenance programmes, and their funding, in a participatory framework.

Approaching ownership and management in this way requires:

- Developing a method to value the road assets
- Developing procedures and processes to serve multiple owners
- Developing procedures for planning in the new administrative environment
- Identifying what kind of government oversight is necessary
- Developing an approach to how the private sector will be represented.

This approach can apply irrespective of whether the asset is being managed by a public or private sector organisation.

2.3 The policy framework

2.3.1 Management and policy

The concept of the *management cycle* has already been introduced. The first step in the cycle is to determine policy and objectives. In essence, this process aims at defining the *policy framework* which provides the context within which decisions can be taken about all aspects of road management. At its simplest, policy may be defined as a 'definitive course of action selected from alternatives to determine present and future decisions' (Howe, 1996). Policies should also be distinguished from plans. Policies are the *what* and *why*; plans are the *how, who, when* and *where*. Policy can be considered to include all of the:

- Relevant laws and statutes, especially their preambles and announcements
- Decision of the courts and regulatory bodies on important issues
- Policy and procedure manuals issued by relevant organisations concerned with the overall management of the road network.

Howe (1996) notes that roads policy should be part of an overall vision for transport that indicates how the supply of infrastructure and vehicles is to be provided, regulated and managed in relation to given demands. Supply policies are determined by a specific, often implicit philosophy defining the role that transport is expected to play in development, and explicit recognition of the types of demands that are being targeted. It should be emphasised that a transport policy should have a vision that extends to all the population, including pedestrians, cyclists and others who are not car owners or users.

A policy framework lays down the basic rules and requirements within which professional and technical decisions can be made about the road network. If this framework is well defined, then there are clear guidelines for determining issues such as the distribution of budgets, about priorities, and all other administrative functions related to the road sub-sector. The role of professionals in the road administration becomes one of providing the appropriate technical solutions to implement the defined policy. Government then needs only to check that policy has been implemented by the professional staff, and to receive feed-back from them about which areas of the policy framework need to be modified so that policy can be improved in the future.

The contents of the policy framework should be set out in a *policy document* that is available for public scrutiny. This should be updated at regular intervals, and can provide the basis of a business plan for the road administration.

2.3.2 Components of the policy framework

A policy framework is normally set at three levels:

- Mission statement
- Objectives
- Standards and intervention levels.

A *mission statement* outlines, in broad terms, the nature of the operation being managed by the organisation responsible for the road network. *Objectives* set specific goals to be achieved within the short- to-medium-term (tactical) and long-term (strategic) time scales. *Standards* and *intervention levels* provide the detailed operational targets to be worked to by individuals in the organisation. Standards may be supported by legislation or regulations; intervention levels may be set internally by the road administration itself. Other terms are sometimes used in this context, and those used here are not intended to be prescriptive. The aim is to give guidance on a simple framework for policy setting.

There should be a 'one-to-many' relationship between the components of the policy framework, as illustrated in Figure 2.1. Each item in the mission statement should be supported by one or more objectives, and each of these should be supported by one or more standards or intervention levels. All standards and intervention levels should support an objective, and all objectives should be reflected in the mission statement. The mission statement, objectives and standards thus provide a consistent set of criteria that can guide all decisions.

When financial resources are plentiful, the absence of a policy framework is not noticeable because management decisions about the allocation of funds are relatively

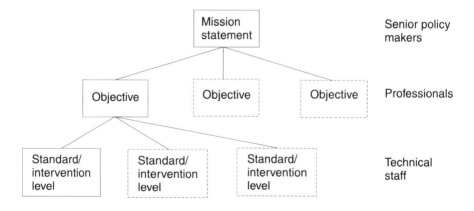

Figure 2.1 *Relationship between components of the policy framework*

straightforward; but when resources are constrained, this absence tends to result in allocations being made to meet short-term crises. This results in funds being allocated in a way that fails to address the longer term needs of the road network; at worst, the situation results in anarchy.

Thus, a policy framework, as defined above, is central to the management cycle. This importance has been recognised by the Commission of the European Communities (1993): their document discusses the causes of low compliance with the objectives of its past lending programme to developing countries. It concludes that, for successful project implementation, the following factors must be present:

- Creation by the recipient country of a rational framework of sectoral policies
- A clear and realistic definition of both the objectives and the recipients
- The drawing of a clear distinction between objectives and the means of achieving them.

The following considerations may assist with policy formulation in the roads sub-sector, and for road maintenance management in particular. They do not attempt to provide a definitive treatise on the subject; rather they discuss some of the issues that should be considered in defining an appropriate policy framework.

2.3.3 Vested interests

One particular reason for separating the policy framework into three levels is that each is applicable to a different audience or vested interests. The principal users of the various components of the policy framework are shown in Figure 2.1.

Mission statements are formulated at a relatively high level, and are normally kept fairly brief, since they are targeted at senior policy-makers and politicians. The mission statement also provides a broad concept to inspire individuals in the organisation, and to enable them to see how their own contribution fits into the wider aims of the organisation. In addition to providing a high level statement of sub-sector policy, mission statements have the purpose of allowing comparison and differentiation of institutional roles and policies between different sectors. Policy is of greatest relevance to the *planning* function of road management.

Objectives are usually targeted at the professional engineers and managers in the administration who have the responsibility for delivering results on behalf of the government. Often these will be the budget holders in the organisation. Objectives enable these results to be quantified in a way that enables achievement against targets to be assessed. Objectives are of greatest relevance to the *programming* function of road management.

Standards and intervention levels provide detailed operational targets to be met on a day-to-day basis. They are normally targeted at technicians, inspectors, supervisors, and others, who have responsibility for ensuring that policy is implemented on the ground. Standards and intervention levels are of greatest relevance to the *preparation* and *operations* functions of road management.

The policy framework provides the basis of communication between personnel at different levels of responsibility within the administration, and enables policy to be implemented in a coherent manner. It is clear that, if this is to be achieved, the components of the policy framework must be defined in a way that is internally consistent (Freeman-Bell and Balkwill, 1993). The relationship between the policy framework and the management functions is illustrated in Table 2.2. It will be seen that the mission statement relates to the network as a whole; objectives relate to that sub-network where work is to be carried out; standards and intervention levels relate to work done on individual sections or sub-sections of road.

The requirements for mission statements, objectives, standards and intervention levels are now described in more detail.

Table 2.2 *Relationship of policy framework to management functions*

Management function	Spatial coverage	Time horizon	Staff concerned	Relevant component of policy framework
Planning	Network-wide	Long term (strategic)	Senior management and policy level	Mission statement
Programming	Network to sub-network	Medium term (tactical)	Middle-level professional/ budget holders	Objectives
Preparation	Section or project (or scheme)	Budget year	Junior professionals	Standards and intervention levels
Operations	Sub-section	Immediate/ very short term	Technicians/ works supervisors	

2.3.4 Mission statement

The mission statement is a recognition of the fundamental purposes for which the road organisation has been set up, and the responsibilities with which it is charged.

Historically, the focus of attention of most road organisations has been on the engineering aspects of the road in terms of the designs and specifications of the pavements and structures. However, the only reason for providing roads is to enable them to be used by traffic. It is, therefore, the needs of the road users that should dictate the policy criteria for the road organisation. It is convenient, for these purposes, to think of the road organisation in terms of a commercial business: the mission statement should then relate to needs of customers and other stakeholders. In this case, the *customers* are the road users, so the mission statement should normally define how the organisation should operate its business to meet their needs.

As such, a mission statement may relate to those areas that were identified in Chapter 1, where impacts would result from works activities on the road network:

- Level of service or road condition
- National development and socio-economic impacts
- Road user costs
- Accident levels and costs
- Environmental degradation
- Road administration costs.

Examples of mission statements that have been adopted in a developing and an industrialised country are shown in Boxes 2.3 and 2.4. An example of a generic mission statement that incorporates all of the suggested components mentioned earlier is given in Box 2.5.

Box 2.3 *Example of a mission statement from India*

The road system will be maintained and operated in a manner such that:

- Comfort, convenience and safety are afforded to the public
- The investment in roads, bridges and appurtenances is preserved
- The aesthetics and compatibility of the road system with the environment is preserved
- The necessary expenditure of resources is accomplished with continuing emphasis on economy.

Source: Ministry of Surface Transport.

Box 2.4 *Example of a mission statement from the United Kingdom*

- To promote an efficient market, with prices reflecting the true cost of transport
- To enable the market to provide greater choice by substantially increasing opportunities for private sector involvement
- By means of the above, to assist economic growth and to provide greater choice in good-quality, safe and accessible services for all transport users
- To promote safety, security and mobility across transport modes
- To reduce the adverse impact of transport on the environment
- To promote UK transport policies and interests internationally
- To promote more efficient, effective and responsive use of Departmental resources
- To conduct licensing, regulatory and grant-payment services effectively and economically, maintaining a high quality of service to industry and the general public.

Source: Department of Transport (1996). Crown Copyright 1996.
Reproduced by permission of the Controller of H. M. Stationery Office.

Box 2.5 *Example of a generic mission statement*

The road administration will aim to provide and maintain a road network to support national development objectives and, in particular, to:

- Provide an acceptable level of service that is economic and safe for road users
- Minimise the sum of road user and road administration costs within the budget available
- Undertake all work effectively, efficiently and safely, and in such a way that minimises damage to the environment.

36 *Road Maintenance Management*

To enable the mission statement to be put into effect, objectives, standards and intervention levels are required.

2.3.5 Objectives

Whereas mission statements are fairly broad in their scope, objectives should aim to set goals and to quantify actions designed to put the policy in place. An objective is the result that a defined policy is designed to achieve. This could be, for example, either to maintain the condition of the road network at a defined standard, or to upgrade the standard of footways in urban areas. To be effective, an objective must be:

- *Measurable*
  Quantified in such a way that it is possible to determine whether or not the objective has been achieved; normally related to achievement within a stated time scale.
- *Relevant*
  Being pertinent and applicable to the administration's mission, and having a direct bearing or influence on the item from the mission statement being considered.
- *Specific*
  Formulated in such a way as to be explicit, distinct and precise, in order to reduce the possibility of misinterpretation.
- *Achievable*
  Such that it is actually possible for the administration to accomplish the requirement in the time available, or defined for response, with the resources available.

Specific objectives need to be set for each area identified in the mission statement; in most cases, several objectives will be needed to cover all facets of the organisation's mission. It is sometimes convenient to group objectives as either physical or operational. Physical objectives define such things as levels of deterioration; operational objectives define aspects such as the frequencies of operation, or response times. For example, the frequency of carrying out *cyclic* maintenance operations may be specified; or the maximum time may be defined for responding to emergencies or, possibly, in dealing with complaints from the public.

Examples of objectives in the different areas of the mission statement are given in Box 2.6. Note that each of the objectives in these examples is likely to meet the criteria of being specific, relevant and achievable, although further definition may be required in terms of being measurable. Where this is necessary, the detailed quantification can be provided through standards or intervention levels.

2.3.6 Standards and intervention levels

Objectives relate to functional areas, such as level of service and user costs. In order to provide criteria for determining whether or not objectives have been met in practical situations, it is often necessary to set standards or intervention levels. The definition of standards and intervention levels differs between road administrations. In some cases, standards are the targets that the road administration should aim to achieve; whereas intervention levels are the minimum level of service that is allowed.

In other cases, standards are defined levels of condition or response that the road administration is obliged by law to achieve in its management activities; intervention levels have a similar purpose, but are norms set by the administration itself. Other terms, such as 'thresholds', 'target levels' and 'warning levels', are also used, and are often set to provide margins of safety or error to the accepted norms.

Box 2.6   *Examples of objectives for different aspects of a mission statement*

Level of service objectives
The road network will be broken down into a functional hierarchy or road classes, each with a defined purpose.

National socio-economic goals
At least 30 per cent of the staff employed by the road administration shall be made up of women, individuals from ethnic minorities or disabled persons.

User cost objectives
Arterial roads will be maintained, so far as budgets will allow, to minimise the sum of user and road maintenance costs in the longer term.

Safety objectives
Road works will be carried out in such a manner as to minimise the hazard caused to road users, pedestrians and workers during the course of the works.

Environmental objectives
Materials used for road construction and maintenance will only be obtained from sources approved by the Ministry of Natural Resources and the Environment.

Road administration cost objectives
The road administration will endeavour to provide value for money by meeting the above objectives at minimum cost, subject to the available budget, and will be able to demonstrate this through third party audit of expenditure on works carried out to meet defined road maintenance standards and assessed need.

Standards may be used to define, for example, geometric designs that are appropriate, or to define thresholds that trigger maintenance actions. In the case of geometric design standards, these are often defined in terms of the required curve standards based on safe stopping or overtaking sight distances; structural pavement maintenance standards are often defined as intervention levels for individual defects, such as the amount of pavement deflection that can be tolerated before strengthening measures are needed.

The following are examples of the formulation of maintenance standards and intervention levels for some of the objectives that are shown in Box 2.4:

- *Safety*
  This requires, for example, that a standard is available for signing at road works.
- *Environment*
  This requires that a standard list of acceptable sources of materials is available.

However, objectives, standards and intervention levels may vary as a function of the nature of the road and the traffic level. There may also need to be variation to suit other circumstances such as terrain or environment. For example, it may be economic to reduce geometric design standards in hilly and mountainous areas. However, although it may be economic to vary standards and intervention levels in different circumstances, it may be politically unacceptable to do so. Clearly, standards and intervention levels need to be set at values that are consistent with the budget level that is likely to be available, otherwise objectives will not be met.

2.3.7 Functional hierarchy

Defining the policy framework demands that customer needs are identified, since a user-orientated approach recognises that the principal purpose of a road is to carry traffic. Different roads in the network serve different purposes, and the 'customers' for these may be different. For example, arterial roads serving national and international centres have the purpose of providing efficient communication links between these centres; the principal customers for these may be transport operators carrying freight over long distances. Since efficiency of operations is important on this type of road, objectives might include the need to eliminate pot-holes so that smooth roads are provided which allow high speeds and reduce vehicle operating costs. By contrast, roads in urban areas have the purpose of supporting directly commercial, industrial, domestic or social activity. The customers for these may be principally the owners of industrial and commercial enterprises for whom roads provide access. The key objective here might be considered to be the requirement for access; high speeds are less important and occasional pot-holes, although undesirable, could be tolerated for short periods, particularly where severe budget constraints exist.

It is therefore convenient to introduce the concept of a *functional hierarchy* to divide the network up into classes, within which roads are subject to common objectives, standards and intervention levels. The hierarchy adopted should be easy to understand and should link directly to the policy framework (Institution of Civil Engineers, 1996b). Defining a hierarchy in this way also has implications for the source of funding for roads of different hierarchies. Arterial roads, which serve a national purpose, could be funded from national sources, whereas the funding for urban roads could be raised and administered totally within the specific urban area. Particularly in times of severe budget constraint, such an approach to funding increases the specificity of operations by clarifying ownership and responsibility, and enables objectives to be set which reflect customer requirements more accurately.

An example of the definition of a road hierarchy in an industrialised country is given in Box 2.7, and an example of that typical in a developing country is given in Box 2.8. An example used for low volume roads is given by the American Society of Civil Engineers (1992).

The Local Authority Associations in the United Kingdom (1989) recommend that road administrations should:

- Categorise their road networks on the basis of an urban/rural split, taking into account the volume and composition of traffic; taking this step will provide a firmer basis for resource allocation in each road administration.

- Use the road hierarchy as the basis for allocating resources and deciding the maintenance priority to be accorded to each class.
- Consider whether it is necessary to grant similar priority to every rural unclassified road serving small communities, and whether some roads should be placed on a minimum maintenance basis.

Box 2.7 *Example of a road hierarchy in an industrialised country*

Category 1: Motorways

Category 2: Strategic routes, including local authority motorways, primary routes and the most important urban traffic links with more than a local significance

Category 3: Distributor roads – both main and secondary, serving a local purpose and connecting to strategic routes

Category 4: Local roads, local interconnecting roads, the remainder of the network

A further subdivision is also used with the above:

Category 3a: Main distributors

Category 3b: Secondary distributors

Category 4a: Local roads

Category 4b: Local access roads

Source: Local Authority Associations (1989).

Box 2.8 *Example of a road hierarchy in a developing country*

Arterial roads
The main routes connecting national and international centres; traffic on them is derived from that generated at urban centres and from the inter-urban areas through the collector and access road systems; trip lengths are likely to be relatively long, and levels of traffic flow and speeds relatively high

Collector roads
These have the function of linking traffic to and from rural areas, either direct to adjacent urban centres, or to the arterial road network; traffic flows and trip lengths will be of an intermediate level

Access roads
The lowest level in the network classification; vehicular flows will be very light and will be aggregated in the collector road network; substantial proportions of total movements are likely to be by non-motorised traffic and pedestrians.

Source: TRRL Overseas Unit (1988c). Crown Copyright 1988.
Reproduced by permission of the Controller of H. M. Stationery Office.

2.3.8 Basis for decision-making

The policy framework provides a consistent mechanism for all decision-making within the road sub-sector. It provides the basis for all actions that are undertaken, and enables these actions to be taken in such a way that customer needs are satisfied and that value for money is obtained. Each policy item should be subjected to an analysis of its costs and resultant benefits in order to meet these ends. This becomes particularly important when there are severe budget constraints, because difficult decisions must be made about how and where money will be spent. Clear benefits result from the use of a rational policy framework, and examples of these are documented elsewhere (Audit Commission, 1988).

Determining the policy framework, through the definition of a mission statement, objectives, standards and intervention levels, assists in the identification and clarification of the aims of the road administration, thereby assisting it to achieve its goals. Policy frameworks should be user-orientated, recognising that the principal purpose of a road is to carry traffic. The structured approach to determining policy is particularly helpful when budgets are severely constrained, because ownership and responsibility are clarified, and objectives can be set that are transparent, equitable and which reflect customer requirements more accurately. The approach also provides a firm basis for planning, by considering options and priorities, determining costs, and monitoring physical achievement and value for money. As such, the concept should be seen as an essential component of the activities of all road administrations.

2.3.9 Performance indicators

The need to be more aware, open and responsive to the customer's expressed needs has led many road administrations to develop performance measures or indicators that they publish for public scrutiny. These need to reflect the policy framework of the administration if they are to be meaningful. In essence, the most important objectives from the policy framework should be highlighted and adopted as key indicators. They need to be reviewed periodically, and action taken on significant deviations from targets (McCoubrey *et al.*, 1995).

Each road administration's performance indicators are drawn up to reflect their own specific policies and directives. As such, they are rarely capable of direct comparison with other administrations, apart from those that are purely financial in origin. Indicators are not necessarily absolute measures in their own right but, by comparing results year-on-year, they contribute to the identification of the relative efficiencies achieved.

Performance indicators can therefore be seen to serve a number of objectives (Humplick and Paterson, 1994):

- *Management decision-making tool*
  To provide inputs for managerial decisions such as investment levels, maintenance expenditures and frequencies, and day-to-day operating decisions, such as traffic management.
- *Diagnostic tool*
  To provide an early warning system that can identify critical locations for investment and allow actions to prevent undesirable outcomes, such as the accelerated deterioration of pavements.

- *Tracking and monitoring tool*
  To assess the adequacy of government and managerial policies by providing effectiveness measures and mechanisms for charting the success of pre-defined policies and objectives.
- *Signalling and audit systems*
  To provide information to users, including communities monitoring the effectiveness of municipal roads and streets, from the suppliers of infrastructure services, and to suppliers from policy-makers in order to guide desired outcomes (such as safety targets at intersections).
- *Resource allocation*
  To support the efficient distribution and control of public resources by quantifying relative efficiency of investments across competing alternatives; for example, performance data can be used as an input to resource allocation ratios across regions and local governments.
- *Information systems*
  To track construction costs and other data of relevance for constructing and managing infrastructure.

The World Bank has suggested that performance indicators should be drawn from those listed in Table 2.3. An example of performance indicators published in a local newspaper

Table 2.3 *Suggestions for performance indicators from the World Bank*

Group	Item
Infrastructure provision	Network size
	Asset value
	Road users
	Demography and macro-economy
	Availability
	Utilisation
Service quality	Road surface
	User safety risk
	Mobility quality
	Accessibility quality
	Road user cost
Provision efficiency	Expenditure productivity
	Output productivity
	Output efficiency
	Provision mode
Sectoral effectiveness	Road function
	Preservation effectiveness
	Road safety
Institutional effectiveness	Resource lag
	Economic return

Source: Humplick and Paterson (1994).

Road Maintenance Management

by a road administration in the United Kingdom is given in Box 2.9. A suggested list of clients for different classes of road, and factors relating to client satisfaction, are given by

---

Box 2.9   *Performance indicators for Bedfordshire County Council in the United Kingdom*

Road lighting

1. (a) The percentage of street lights not working as planned — 1.2%

    (b) The percentage of street lights not working as planned because they are awaiting repair from the Road Administration — 1.1%

    (c) The method of inspection used to monitor the above — Regular night-time inspections of ALL street lights

Maintaining roads and pavements

2. (a) The Administration's definition of damage to roads that will be repaired or made safe within 24 hours — "A hole in the bituminous surface with approx. vertical sides, where material has been lost, and where surface dimension in two directions exceed 150 mm and depth exceeds 50 mm"

    (b) The percentage achievement within 24 hours of repairs and/or damage made safe in this category — 83%

    (c) The Administration's definition of damage to pavements that will be repaired or made good within 24 hours — "A hole in the bituminous surface with approx. vertical sides, where material has been lost, and where any surface dimension in two directions exceed 100 mm and depth exceeds 20 mm
    or
    a difference in vertical level between adjacent slabs and other projections, such as man-holes, boxes and the like exceeding 20 mm; cracks of gaps between adjacent slabs, or between adjacent slabs and other projections, such as man-holes, boxes and the like, exceeding 20 mm^2"

    (d) The percentage achievement within 24 hours of repairs and/or damage made safe within this category — 89%

3. The cost of highway maintenance per 100 miles travelled by a vehicle — £0.41

Providing pedestrian crossings

4. The percentage of pedestrian crossings with facilities for disabled people — 49%

Source: *Luton & Dunstable Herald & Post*, 28 December 1995. © Bedfordshire County Council – Department of Environmental and Economic Development

Haas and Hudson (1996). Other examples of performance indicators, and methods of their monitoring, are given by the Transportation Research Board (1994).

2.4 Organisations

2.4.1 Organisational culture

Definition of a policy framework also provides a firm basis for determining organisational structures that are designed to meet customer requirements, and deliver the mission of the road administration. Different organisational forms tend to have different cultures (Handy, 1993), and four of these are summarised in Table 2.4.

Table 2.4 *Organisational cultures*

Type	Characteristics
Power culture (a 'web')	• Characteristic of a new or a family business • Central source of power • Trust and empathy between people • Communication by personal conversations • Few rules and procedures; little bureaucracy • Control exercised by the centre through key individuals • Organisation can react and respond quickly to change • Individuals are judged by results, and organisation is tolerant of the means of achieving these
Role culture (bureaucracy)	• Characterised by organisational separation by function (e.g. finance, procurement, operations, etc.) • Interaction between departments is controlled by procedures for roles (including job descriptions, authority definitions, etc.); procedures for communications (including filing systems, required copies of documents, etc.); rules for the settlement of disputes (e.g. appeal to the lowest cross-over point in the structure) • Co-ordination is by a small band of senior managers • Roles are more important than the person filling the position, and a wide range of individuals can fill positions; performance over and above the requirements of the role is not required, and can be disruptive • Individuals have security and a predictable way to progress; personal power is unwelcome; expert power is tolerated only in certain positions • Organisation is slow to perceive need for change; and slow to respond to change, even when need is perceived • Useful where economies of scale are more important than flexibility, and where depth of experience is more important than innovation

(continued)

Table 2.4 (*contd.*)

Type	Characteristics
Task culture ('matrix')	• Orientated towards a project or job; bringing together appropriate resources to get the job done • Adaptable organisation, with teams formed for specific tasks and being disbanded afterwards • Influence is based more on expert power than on position or personal power • Control in the organisation is difficult – essentially done through the allocation of resources • Appropriate where flexibility and sensitivity to market situations are important; in very competitive situations, or where speed of reaction needs to be quick, or project life is relatively short • Organisations flourish where business climate is agreeable, where the end result is all-important, and where customer is always right; but problems arise when money is tight and resources have to be rationed
Person culture ('cluster')	• Organisation is effectively a collection of highly expert individuals • Minimal structure and control mechanisms • Such organisations normally evolve into a power culture or a task culture

Adapted from: Handy (1993). Reproduced by permission of Frederic Warne & Co.

The requirements of a road administration are unlikely to be met by a 'person culture' type of organisation, which is more suited to a small civil engineering consultancy. But aspects of each of the other three organisational cultures may be appropriate. A typical road organisation would be expected to embrace the following work areas.

Policy

- Overall direction and management of the organisation, usually at board or director level.

Road management

- Planning, programming, preparation and operation of routine maintenance, periodic maintenance and development
- Operation of special works of emergency and winter maintenance
- Administration, personnel, finance and legal support.

Organisational development

- Research and development
- Training.

These work areas could be carried out by different organisational types as shown in Table 2.5 and as illustrated in Figure 2.2.

Institutional Issues 45

Table 2.5 *Organisational requirements*

Organisational function	Appropriate culture	Notes
Policy	Power	Requirements for direction and leadership; for being results-orientated; and for developing the business from a new organisational base
Road management: • Routine and periodic maintenance, development • Administration, personnel, finance and legal support	Role	Requirements for undertaking relatively routine and programmable activities, with emphasis on quality; relatively large size, with some economies of scale
Road management: • Emergency and winter maintenance operations	Power	Requirements for very rapid response to non-continuous and discrete operations; needs to be flexible and adaptable; results-orientated
Research & development/ Training	Task	Requirements for multi-disciplinary working; possible use of high and rapidly changing technology

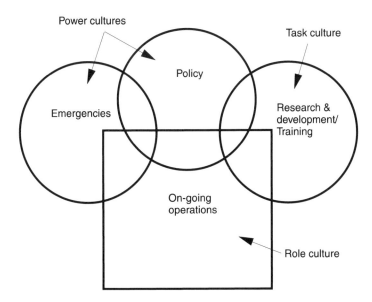

Figure 2.2 *Organisational model (Adapted from*: Handy, 1993). Reproduced by permission of Frederic Warne & Co.

2.4.2 Size and degree of decentralisation

Size

As noted in sub-section 2.1.3, many countries are moving towards a more market-based approach to road management, and this has particular consequences for the way that organisations should be set up. Such an approach is likely to mean that management of the network can be handled by fewer staff covering only the broad planning and management functions listed above. The regular staff would be supported by contractors for maintenance development works. The resulting skill mix of the professionals in such a road administration clearly differs from that in the broad spectrum of organisation that has typically existed in the past. Separation of functions in this manner is discussed in more detail in section 2.5.

An example of an organisational structure is shown in Box 2.10. The size of a road administration, in terms of the number and skills of staff employed, will need to vary depending on the particular characteristics of the country concerned and the network to be managed. There are no generally accepted norms for this, but experience suggests that the size of network that can be managed efficiently is likely to be in the range of 500–3000 km (Schliesser and Bull, 1993). It will be seen that this is consistent with the responsibilities of the regional offices in the example in Box 2.10. In densely trafficked urban areas, the appropriate figure is likely to be at the lower end of the spectrum, and more lightly trafficked rural networks are likely to be at the higher end. This size of operation has the following advantages:

- It is large enough to employ a critical mass of professional and technical skills
- It can include a length of road that is cost-effective to manage as an entity
- It is small and regional enough to reflect local needs and opinion.

Decentralisation

The appropriate degree of decentralisation will depend on a number of factors. These will include:

- The funding and physical responsibility (ownership)
- The policy framework
- The diversity of the road hierarchy or classification
- The geographic distribution of the road network.

The diversity of these factors will differ from country to country and, possibly, within a country. For example:

- The type of road user on different parts of the network may be different.
- The main or trunk road network may be used predominantly by long-distance traffic who have a need for relatively fast and efficient transport.
- The purpose of local roads may be to provide access to industrial or commercial premises, so the need is primarily for access rather than speed and efficiency.

| Box 2.10 | Example of staff requirements in a medium-sized road administration |

The World Bank have suggested the following as the typical requirements for a road administration where all work other than emergency works is contracted out. For an organisation maintaining a road network of 20 000 km, of which 5000 km is paved, there would be as requirement for 800–1000 regular staff, consisting of 40–50 engineers, about 100 technicians, administrators and foremen, and 500–600 labourers. It would also require 5000–6000 sub-contract staff for carrying out routine maintenance, and a further 2000 staff would need to be employed by contractors for rehabilitation and periodic maintenance work. This is illustrated below.

Function[a]	Managerial/ professional	Technical/ admin/ foreman	Charge-hand/ supervisor	Labourer
Management	5	10	—	—
Other HQ staff	15	25	—	—
Regional offices	20	50	500	—
Road management system	3	10	10	—
Mobile maintenance unit[b]	2	6	12	120
Training production unit[c]	2	5	25	—
Total	47	106	547	120

Notes:
a 20 000 km road network managed through ten regional offices.
b Mainly for emergency maintenance.
c Provides on-the-job training in all maintenance activities.
These figures were originally proposed for typical road administrations in Africa; it should be noted that the actual staff numbers in road administrations in countries such as Finland and New Zealand are significantly lower than these.

Source: Heggie (1995).

As a result of this diversity of factors, standards and intervention levels should be different for different classes of road in the hierarchy to reflect the different purposes of travel. Also, there may be different sources of funding and responsibility for managing roads in different classes, or in different regions. There are also interrelationships between these factors. The following are examples of these:

- Maintenance policy would be likely to reflect the needs of different road users and these, in turn, would be reflected in the road hierarchy.
- Objectives, standards and intervention levels that are set should relate to the availability of budget to enable them to be met.
- The source of funds and the budget allocation are also likely to differ for different levels in the hierarchy, and all of these may depend on the organisation responsible for road management in a particular geographic area.

The overlapping and interrelating nature of these factors can be illustrated conceptually as in Figure 2.3. Each of these factors need to be taken into account when deciding on the appropriate degree of decentralisation for road management activities. An example of the issues to be considered in a road network where the management of different classes of roads is decentralised into three groups is illustrated in Table 2.6.

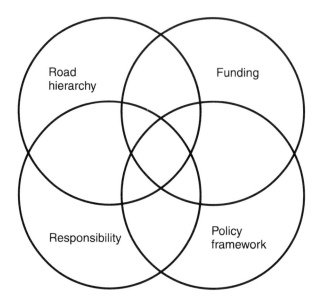

Figure 2.3 *Conceptual interrelationships between factors affecting appropriate degree of decentralisation*

An important consideration is that all of these factors, and the relationships between them, will change over *time*. For example, the physical network may change; as may policies and priorities; traffic volumes may grow and vehicle types may change on particular links requiring changes in hierarchy; budget levels will certainly change; and responsibilities may change as regional or local government is reorganised, or as contracts for the management or execution of works are won by different firms. The appropriate degree of decentralisation may change over time as the needs of the network change.

In addition to public roads, other types of ownership may also exist:

- Privately owned and funded roads, such as within industrial or agricultural facilities, or on private land or estates.
- Privately owned roads, for which there is a government contribution of funds or subsidy, such as forest roads which may be open to the public for leisure pursuits.
- Concessions, which are normally funded from toll revenues.

These all represent other types of sub-networks or road links that are a part of the national road network, and the same considerations for decentralisation also apply.

Table 2.6 *Example of the issues to be considered for decentralisation*

	Road classification		
	Primary roads	*Secondary roads*	*Tertiary roads*
Road hierarchy	Main, strategic or trunk roads will serve needs defined for the whole network	Secondary roads may have a function related to the geographic region within which the sub-network lies	Tertiary roads may serve a purely local function
Funding	Fund allocations for roads serving network needs are likely to come from central government sources	Funding for a geographic sub-network will often be raised largely at the regional level in which the sub-network lies	Funding for road links in the tertiary network may often be raised largely at local level
Physical responsibility	Central government will normally have the responsibility for works on roads serving network needs, although the actual works may be managed under contract or on an agency basis	Responsibility for works on a sub-network will normally reside with the region in which the sub-network lies, although works management may be sub-let	Local government will normally be responsible for managing individual road links in the tertiary network
Policy framework	Objectives, standards and intervention levels will reflect mainly national needs, and be set nationally	Objectives, standards and intervention levels will reflect mainly regional needs, and be set at this level	Objectives, standards and intervention levels will reflect mainly local needs, and be set locally

2.4.3 *Organisational status*

If the administration of roads is separated from government, an important decision has to be taken about where this organisation sits in the public–private domain. The following options are likely to be viable:

- Government department
- Road authority
- Road agency.

The option of the road administration being a department of government has been discussed above. Such an arrangement does not have the benefit of functional separation, and lacks specificity (see sub-section 2.5.2). This is not recommended.

Conceptually, a road authority is a public sector body that reports to a roads board. To be effective, the roads board should be drawn from the stakeholders in the road network, and have majority membership of private sector road users. It needs to have freedom to set the road tariff (see Chapter 3), to determine its own business plan and management arrangements, and to offer competitive terms of remuneration to attract and retain competent staff. Note that a 'parastatal' that has a board comprising representatives of government, and whose staff conditions are determined by government, does not constitute an 'authority' under this definition; it is, in effect, a government department.

A road agency is a private sector company that is awarded a contract under competitive procurement to manage the road network on behalf of government. This would normally be for a fixed term. (Note that, in North America, the terms 'authority' and 'agency' are often used in the opposite sense from that defined here; caution should be exercised when interpreting documents using these terms to ensure that the precise meaning is understood).

2.5 Effectiveness and efficiency

2.5.1 Basic concepts

A key aim of a road administration should be to increase effectiveness and efficiency. *Effectiveness* measures the capability of an organisation to define, agree and meet appropriate operational objectives. *Efficiency* refers to the ability of an organisation to meet those objectives (i.e. to be effective) using minimum resources. Research undertaken by the World Bank (Israel, 1987) in a number of sectors identified that key requirements for achievement of these aims were to increase the *specificity* of an organisation and to subject it to *competition*.

2.5.2 Specificity

This is the feature of an organisation that enables it to identify and focus on specific objectives, without being side-tracked to unproductive tasks. Management structures and procedures need to be designed specifically to achieve these objectives, and well-defined objectives focus the providers on increasing efficiency while, at the same time, meeting customer demand. Several items may be considered in order to raise specificity:

- *Objectives*
  should be set in terms of output, and defined with as much precision as possible.
- *Time*
  periods for meeting objectives should also be defined closely; longer time periods usually imply lower specificity.
- *Procedures*
  for achieving objectives should be carefully defined to ensure easy measurement of performance and efficiency.

- *Control*
  of achievement or output is required by collecting data and information in order that accomplishment can be verified against objectives and methods.

In particular, specificity can be increased by addressing the following issues:

- Identification of the stakeholders in the services being provided
- Definition of the policy framework
- Functional separation to increase focus on activities
- Introduction of commercial management practices.

Stakeholders

Stakeholders were listed in sub-section 2.1.4. Specificity is increased by gearing services to the needs of identified customers. This can be increased further by involving formally all stakeholders in the planning and management functions related to the road network.

Policy framework

Clear definition of the policy framework increases the specificity of an organisation. The issues involved were discussed in section 2.3.

Functional separation

The concepts involved in 'functional separation' will be discussed in sub-section 2.5.4.

Commercial management practices

Increasing specificity by introducing commercial management practices includes the need for (Heggie, 1995):

- Introduction of appropriate management structures and operational procedures
- Strengthening of managerial accountability
- Addressing human resource requirements, including employing sufficient (and only sufficient) staff with adequate skills
- Improving access to information by implementing management systems in the areas of
  - finance and accounting
  - road management
  - bridge management
  - personnel management.

Such commercial management practices increase specificity and provide a framework within which competition can take place.

2.5.3 Competition

Competitive pressure is also a mechanism for increasing effectiveness and efficiency. Cox (1987) has described how competition introduced into public sector maintenance organisations in the United Kingdom resulted in dramatic efficiency increases in these areas. In this sense, competition is defined more broadly than in conventional economics: in addition to external competition from others, competitive pressures can be exerted on an organisation by the political establishment, regulatory agencies and by road users, and by managerial measures that create a competitive atmosphere within the organisation (Israel, 1987). Competition provides users with choices that can improve the way that their needs are met, and that compel providers to become more efficient and accountable; competition can be:

- *External*
  Such as between private contractors.
- *Internal*
  As can sometimes be possible between different departments of an organisation.
- *Mixed*
  Where a public sector organisation competes with organisations from the private sector.

An example of the last of these may be found in the United Kingdom, where road administrations must compete with the private sector for road management and execution of works (Madelin, 1994a). Dramatic increases in efficiency have resulted. Some public sector bodies, who have failed to compete effectively, have been closed down by government.

It should be noted that road works undertaken by contract are only likely to be performed more effectively than those carried out by a government department if the contractor is subject to competition. The main issue is the degree of monopolistic control exercised, rather than whether there is public or private sector involvement in the works. Note that a parastatal or private body operating in a monopoly position has little incentive to perform better than a government organisation, and both can be much less accountable in terms of price and level of service. It is, therefore, emphasised that the key to effectiveness and efficiency is competition; not merely *privatisation*.

The execution of new works and periodic maintenance can readily be performed by the private sector under contract, and this is done routinely in many countries. However, undertaking road maintenance works by contract raises certain issues. A World Bank report on this subject (Miguel and Condron, 1991) reached the following conclusions:

- The practice of contracting road maintenance works is spreading.
- The primary reason for shifting from in-house works account to contractors was to improve effectiveness, but the decision was made without prior consideration of cost comparison.
- Contracting out improves transparency and accountability, and obliges carrying out a thorough preparation, which facilitates good performance and supervision.

- It is not easy to compare costs when both systems are in place and both are efficient (tax regimes, for instance, are not identical) . . . it is accepted that in-house units work better when they have to compete with private contractors, and particularly if quality control is done by independent controllers.
- Road maintenance works are usually paid for by using measured work contracts; there are also performance contracts, but they are more sophisticated to set up.
- Maintenance operations are packaged for contracting purposes according to the type of works with little distinction made between 'routine' and 'periodic' maintenance; this distinction is made mainly for budgetary purposes.
- Long-term commitment to contract maintenance programmes must be made by road administrations to encourage interest from the private sector; a smooth transition requires planning and a close relationship with the contracting industry.

Experience of undertaking road maintenance works by contract in several countries around the world suggests that cost savings are typically in the range 10–15 per cent. Some countries are also undertaking planning and management functions by contract, and similar savings have been reported for these (Boddy, 1988).

The use of contractors is discussed in more detail in Chapter 9.

2.5.4 Functional separation

Government and administration

The separation of the roles of government from that of the road administration was discussed in sub-section 2.1.3. Recognising that the role of government is to define policy, legislation and to make budgets available, it then becomes clear that putting policy into effect through administration can be separated functionally from government. Once the detailed operational rules have been defined in the policy framework, then any organisation can be tasked with managing the asset concerned by putting these rules into effect. In essence, this means meeting the objectives of the policy framework.

The road administration, in principle, can be a public, parastatal or private organisation. It has the role of executing the road programme to meet the aims of government, setting detailed rules and standards, maintaining, developing and, more generally, managing the road network. In some cases, the road administration has the responsibility for enforcing regulations for such things as traffic safety and axle loading. By monitoring performance of the road network, the road administration can advise government on changes to policy that may be advisable to improve the way that the network is managed in the future.

Client and supplier

The functions of a road administration can be split conveniently into those for the planning and management of road operations (the 'client' function), and those for works execution (the 'supplier' function). The client role is concerned with specifying activities to be carried out, determining appropriate standards to use, commissioning works, supervising, controlling and monitoring activities. The supplier role is concerned with delivering the defined product to an agreed quality standard, to time and to budget.

Such separation increases the specificity of both the management and the works execution functions. It enables organisations to operate more as a business than a bureaucracy, and it helps to clarify missions and objectives. This enables operations to be run as a service industry that responds to customer demand.

The management functions of a road administration are shown in Table 2.2. In addition to these, there will also be development functions of research and training, plus functions of administration, personnel, finance and legal activities, as in Table 2.4. Each of these functions can, if required, be broken down into client and supplier activities. There are four main options for the provision of services in this manner:

- Public ownership, with operation by a department or a parastatal
- Public ownership, with operation contracted to the private sector
- Private ownership and operation that is subject to regulation
- Community provision.

Consideration of these various factors leads to a generic organisational structure for the direction and managerial levels of a road administration, and this is shown in Figure 2.4. There will be further subdivision into activities for some of these functions.

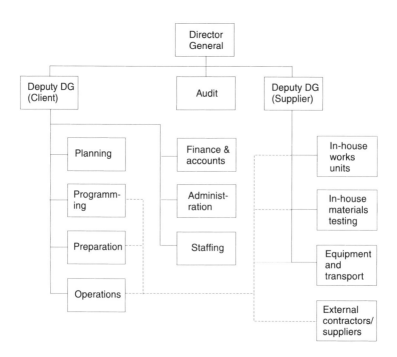

Figure 2.4 *Organisation with functional separation of client and supplier*

Organisational reforms in the road sub-sector have found that separating the client and supplier functions increases the focus of action of individuals in the road administration. Greater benefits in effectiveness and efficiency are obtained from this separation than from virtually any other type of organisational reform (Madelin, 1994a). There are, as a result, good reasons why functional separation along these lines should be a fundamental consideration in any restructuring of road administrations.

2.5.5 Commercialisation

Many road administrations in different countries are seeking to transform their operations to work in a more business-like manner and to become more effective and efficient. 'Commercialisation' is one approach to achieving this. Commercialisation does not necessarily involve 'privatisation', and the terms should not be confused. There are three main steps in the commercialisation process where the aim is ultimately to achieve privatisation, at least of the 'supplier' functions. Table 2.7 illustrates these and gives the terminology that can be used in association with the process.

Table 2.7 *The commercialisation process and terminology*

Steps	Corporatisation option	Joint venture option
Step 1 Functional separation	Separation of organisation into 'client' and 'supplier' functions, and other internal measures to improve effectiveness and efficiency (as described in section 2.5, for example)	
Step 2 Divestiture	Corporatisation of parts of the public sector organisation by transformation into parastatal or joint stock companies, with government retaining majority or total ownership	Establishment of joint ventures between parts of the public sector organisation and local or foreign private firms, normally with government having a minority ownership
Step 3 Privatisation	Organisation owned 100 per cent by the private sector	

Different mechanisms of commercialisation can be adopted, including divesting parts of the existing organisations as functioning entities, or seeking joint venture arrangements with contractors. A phased approach to commercialisation can be adopted such that organisations will receive declining support over time from the roads administration in terms of the value of the work that is guaranteed. This provides protection to the roads administration in the medium term to ensure that there is capability to carry out its maintenance programme. It also gives protection to the works organisation by allowing time to develop its business skills before complete divestiture. Further protection is given to the client because privatisation need not take place until the market is sufficiently strong to avoid the problem of cartels being formed by a small number of qualified contractors.

Any commercialisation plan should be flexible. Short-term actions need to be defined in considerable detail, whereas the detail of medium or longer-term objectives can be less precise. This allows modification following each stage of the process to enable actual experience and difficulties encountered to be reflected in the approach.

2.6 Institutional development

2.6.1 Hierarchy of management issues

A study of institutional development in several road organisations identified three main areas which need to be considered when determining the capability of an organisation to undertake road management effectively and efficiently (Brooks *et al.*, 1989). This study identified an interdependence between the external environment in which the organisation operates, the internal institutional arrangements of the organisation and the technical capability of the organisation. External factors, and internal institutional and technical factors, were discussed in sub-section 2.1.1. The study concluded that effective performance in one area cannot be achieved without complementary performance in other areas.

The study suggested that failure to make sustainable improvements to road management practices in the past has been because efforts have generally been focused on improving the technical capability of organisations, without first ensuring that internal or institutional capability existed to support this, and that the external context was conducive to effective and efficient management. Improvements in these aspects are normally prerequisites to improvement in technical capability. The study postulated that a hierarchical relationship existed between these factors. This hierarchy can be considered as a pyramid, with external factors as the foundation, technical factors at the top, and institutional factors in the middle. Developing institutions requires that an appropriate external environment needs to exist before institutional factors can be addressed in a satisfactory manner, and that unless there is sufficient institutional capacity, then it is not possible to develop technical capability. This pyramid is illustrated in Figure 2.5.

This indicates that the pyramid must be built from the bottom up. For technical improvements to be successful, sufficient institutional capability must exist. Furthermore, unless the external context is appropriate, the whole pyramid will collapse, and no sustainable improvements can be made.

2.6.2 The challenge of change

As the national economies of many countries move more towards a more market-driven approach, so road administrations will need to change to reflect the different demands that are placed upon them. These demands will reflect increased pressure from customers for an improved level of service, and for greater effectiveness and efficiency of operations. Such changes are likely to be painful; people have a natural reluctance to change. Any change must be handled carefully and needs to be managed pro-actively.

Institutional Issues 57

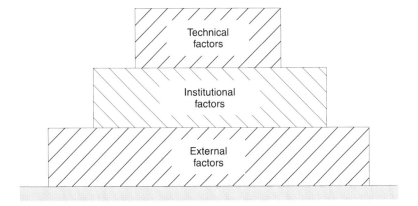

Figure 2.5 *Hierarchy of management issues (Adapted from:* Brooks *et al.*, 1989)

A logical and structured approach to change should reflect the pyramid approach to organisational factors discussed earlier and illustrated in Figure 2.5. This proposed that there was a need to address external factors before institutional factors, and that technical factors should in general only be addressed after changes have been implemented in these other areas. Since the pyramid must be built from the bottom up, this suggests the following general order for implementing change:

1. Obtain political and governmental commitment to change; without this, real change is unlikely to be effective – involvement of users and other stakeholders can help to obtain this commitment.
2. Agree the policy and institutional framework; questions need answering on issues such as: who will be responsible for different parts of the road network?
3. Frame any legislation needed to support changes where this does not already exist.
4. Secure funds; these need to be sufficient for the implementation of policy, and need to be available on a regular and stable basis.
5. Re-cast organisations to undertake the work, with consideration of how best to achieve effectiveness and efficiency.
6. Technical issues should be considered when the above external and internal institutional factors have been addressed.

This order is not intended to be prescriptive, but only suggests a general approach to the tasks to be undertaken. For example, there may well be factors that can be addressed without legislation being passed, and there is no reason why these should not be undertaken early if the opportunity exists; there may be technical standards that can be improved relatively easily and cheaply, which could have a beneficial impact without organisational changes having been put into effect. The key point is that a structured approach to addressing institutional change is necessary in which strategic goals are set, and within which short-term actions can be taken to move along the transitional path.

Obtaining political commitment to any reform is the first step that must be undertaken. This is not usually easy. Heggie (1991a) has identified the following actions that can be undertaken to facilitate the achievement of this:

- Identify the proposed reforms, and the potential political and organisational constraints on reform, on the basis of a detailed institutional analysis.
- Be sure that government considers reforms timely and relevant, and that there is a committed constituency within government willing and able to implement them.
- Estimate the benefits and costs of reform.
- Be sure that government has the administrative resources to implement the reform.
- Include special measures aimed at overcoming political and organisational constraints.
- Establish a reasonable timetable.

Heggie (1991a) also notes that major policy reform takes 10–15 years to complete, and should therefore be seen as a long-term process.

2.7 Human resource management

2.7.1 Human resource needs in a changing environment

As noted earlier, many road organisations are undergoing substantial change. Any change will affect people; and their needs should not be forgotten. People will be needed to instigate change, and to carry it through to its conclusion. In many areas there will be a need for changed attitudes to all aspects of work; often these attitudes will have been entrenched for many years, or even generations. There is a need for ownership of change, and to assign clear responsibility and accountability for all actions associated with change.

It will be appreciated that the types of organisations, such as implied here, will have a different skill mix than many existing road administrations. They will be smaller, more commercial, and reflect only the client functions. As a result, they will need more managers, more staff with financial backgrounds, and more engineers with experience of contract management, contract law and arbitration procedures. Such changes require new personnel policies and great emphasis on training programmes: the needs for this should not be underestimated. Similarly, there will be a need for the development of contracting organisations operating on commercial lines, again with a significant need for corporate and personnel development.

The human resource aspect of the transition process offers a critical trial which epitomises the wider challenges that are present. Addressing this carefully and sensitively will do much to smooth the institutional transition where it is now under way.

2.7.2 Skills and motivation

Where personnel have inadequate skills or lack motivation, the effectiveness and efficiency of operations are severely constrained. Training has often been seen as a panacea for solving problems in this area, but the record of training in many situations has been disappointing. One reason for this is that, too often, training has been seen in isolation from the broader subject of human resource development. Sometimes, too little attention has been paid to manpower analysis before training is planned, or to manpower management after training has been completed. Insufficient analysis of both manpower resources and needs carries the risk that the wrong type of training may be given, or that people are trained for the wrong jobs. Poor manpower management means that trained staff are unable to apply effectively what they have learned, with the consequence that their training is wasted (Relf and Thriscutt, 1991). Whereas the earlier focus was on training, it is now becoming apparent that the human resource problem is more a question of utilisation, motivation, development and retention (Robinson, 1991b).

There are a variety of factors which may cause ineffective use of manpower, including poor personnel policies resulting from the application of inappropriate rules of service, conditions of employment and pay, and lack of accountability and incentives, the level of an organisation's efficiency and its structural complexity. These issues tend to be a particular problem within public sector agencies. The human resource problems seem to be exacerbated when considering road maintenance because of other institutional constraints inherent in the maintenance process.

The development of human resources frequently can take place much more effectively when the organisation is able to operate in an autonomous or semi-autonomous manner, and where it can hire only those people that it actually requires to do the job. Such an organisation also has more flexibility to pay people at a level required to retain the talent it needs. Adequate staff remuneration is a key component of an effective road administration. However, public sector salaries are often below those paid in the private sector, and civil service conditions often constrain operations. Thus, commercialisation of road maintenance management can assist in improving effectiveness and efficiency.

Even without autonomy, an organisation can still institute innovative compensation schemes, such as remuneration that is linked directly to quantities and qualities of outputs. Such a compensation system is an effective tool for strengthening manpower management. Effective manpower utilisation also requires a good system of job descriptions, management by objectives, appropriate workload allocations, and effective staff supervision. These are basic personnel and work management tools.

2.7.3 Training

Purpose of training

The aim of training is to improve job performance by extending knowledge, inculcating skills and modifying attitudes, so that individuals can work in the most economical, efficient and satisfying way (Barber, 1968). Training is not optional: it is essential to both organisational and staff development. Without training, stagnation results. In the context of road organisations, the focus on training should be to ensure that an organisational culture is developed, to enhance motivation, co-operation, communication and to facilitate the use of the management systems being introduced. Thus, training is a way of improving performance by changing the way that work is done. It can be used to assist the development of:

- Management techniques
- Work methods and practices
- Technical skills.

There is normally a particular need for the training of road maintenance staff. Often skills have been learned on-the-job, and staff have a very practical approach. There is often a lack of exposure to theoretical concepts, with the result that a very conservative attitude prevails with respect to change. Training programmes are also important for creating or reinforcing the ability to undertake environmental assessments. Training programmes are most effective if scheduled well in advance. They should also be adapted to the tasks of the personnel involved.

The different types of training for the roads sub-sector have been described by Thagesen (1996).

Conditions for success

Training to support organisational development needs to be formulated with considerable care if sustainable results are to be obtained. In particular, training should always be designed to meet clear objectives, with achievement that is measurable. Hence, emphasis should be placed on the outputs from training rather than the inputs in terms of the number of people trained. Training needs to be developed in the context of the organisation's wider aims and objectives. As such, there should be a high degree of correlation between training issues and institutional issues. Experience has shown that an institutional appraisal, covering external, institutional and technical issues, is the most effective method of identifying clearly training objectives together with institutional constraints.

Training needs analysis

In order to identify training needs, an analysis should be undertaken to include consideration of:

- Institutional issues
- Training environment
- Current performance levels
- Required performance levels
- Gap in performance levels to be rectified by training.

The training needs analysis is likely to identify more training than can be afforded, and decisions must be made about which to undertake now and which to defer. The aim should be to set training priorities on the basis of cost-effectiveness: the choice of training to be undertaken should be that which gives the greatest return on investment. All potential training should be evaluated in this way (Thagesen, 1996).

2.8 Environmental issues

2.8.1 Issues of concern

Roads can have a significant impact on the environment and on the lives of people (World Bank, 1994). However, whereas the provision of new transport infrastructure may have a significant impact on the physical environment, the impact of maintenance tends to be much less severe. The challenge for the road manager is to avoid, mitigate or compensate for environmental problems when planning and undertaking road works. Wherever possible, ways should be sought to improve the environment surrounding the road. It is preferable to identify potential problems at an early stage, otherwise unsatisfactory, and often expensive and time-consuming, compromise solutions may have to be adopted.

For management activities on existing roads, impacts on the physical environment in the following areas can be anticipated:

- Consequential developments that utilise transport
- Depletion of material resources
- Erosion of soil
- Ecological damage
- Hazardous materials
- Air pollution
- Nuisance due to dust.

Management activities should also attempt to address impacts on the human or social environment in the surrounding area, including:

- Community life, cultural heritage and economic activities, including changing demand on health services and education
- Land acquisition and resettlement
- Noise and vibration.

A detailed discussion of environmental issues is beyond the scope of this book.

2.8.2 Environmental assessment

Assessment

Environmental assessment of road works should not be seen as a single activity at one point in time, but as an on-going process (OECD, 1984) integrated with the management functions of *planning, programming, preparation* and *operations*. Different types of assessment are generally appropriate for different functions, as shown in Table 2.8.

Project screening

The principal purposes of project screening are:

- To ensure that environmental issues are given sufficient attention
- To indicate the level of environmental assessment that is needed.

Screening creates an awareness of significant environmental impacts at an early stage, and helps to ensure that resources are spent in the most effective way.

Table 2.8 *Environmental assessment*

Management function	Environmental assessment method
Planning	Screening
Programming	Preliminary environmental assessment
Preparation	Environmental impact assessment
Operations	(Review and monitoring)

Preliminary environmental assessment

These can be undertaken for works not requiring a full environmental impact assessment (EIA). Typically, the following questions will be posed:

- What are the likely effects of the works?
- Who will be affected?
- What impact will be there be on the physical environment?
- What mitigation measures can be adopted?

The implications of not carrying out the works (that is of doing nothing) should also be considered.

Environmental impact assessment

This seeks to determine the most likely impacts so that mitigation measures may be taken. It normally involves the following tasks:

- Identification and comparison of different alternatives
- Scoping, in terms of range of impacts, affected area and timescales, to assist in focusing on specific impacts
- Prediction of size and nature of impacts
- Evaluation of alternative approaches to the works
- Identification of mitigation measures
- Presentation of results.

Results

Where environmental impact statements are required, these can be presented in the form of a 'balance sheet', where all costs and benefits are set out using monetary or physical units, where possible, or a points rating where not. The statement should also indicate who are the bearers of costs and the recipients of benefits. For unquantifiable environmental effects, the following general principles should be considered:

- The population exposed to the effect should be enumerated and described in terms of its level of sensitivity to the effect, recognising that levels of sensitivity may vary for different population groups.
- The existing level of the effect should be identified and measured, where possible, to show the current degree of exposure of the population.
- The anticipated level of the effect resulting from the works should then be predicted, in the same units as for the existing level, to enable comparisons to be made in terms of population affected.

Consideration should be given to the reliability of information on the local social environment, site conditions and climate. It may be necessary to obtain additional data during the preparation of the works to increase the reliability to an acceptable level. Alternatively, it may be possible to formulate the works in such a way to reduce the level of risk.

3 Finance

- Allocation of funds
  - the budgetary process and budget heads
  - allocation and disbursement mechanisms
- Revenue
  - principles of pricing and cost recovery
  - sources of revenue
  - financing road maintenance and development
  - managing the revenues
- Road funds
- Congestion management
- Road tolls
  - advantages and disadvantages
  - pricing the toll
  - investment and design criteria
- Private financing
  - advantages and disadvantages
  - the parties involved
  - contractual framework and risks
  - shadow tolls
- Local financing

3.1 Issues

Effective management of road networks requires that budget levels are at least sufficient to keep the core road assets in stable condition in the long term. This requires that on-going maintenance is funded, and that adequate provision is made for any strengthening works required. More than this minimum level will be required if the network is to be expanded or improved. Funding for roads traditionally comes from governments, but there is a growing trend to seek other sources of finance for building and managing networks. In addition, many governments are seeking improved instruments for funding roads as part of their efforts to restructure the road sub-sector and make it more commercial.

Many countries find it difficult to maintain road funding at past levels. Part of the reason for this is fiscal pressure on general government revenues, but there are also some more fundamental reasons why the need for adequate road funding is not always seen as a priority by governments (Heggie, 1995). Typically, roads are seen as a 'common good', funded like a social service. Road users pay taxes and road user charges, and the proceeds are nearly always treated as general revenue. Roads are funded through budget allocations determined as part of the annual government budgetary process. These allocations often bear little relationship to the levels of funding that road users actually contribute to revenue or to underlying needs of the network, measured in terms of economic criteria. Consequently, there is no direct linkage between revenue and expenditure.

This mechanism of funding means that users generally do not perceive any *price* for road use. As a result, there tends to be bias in the way that roads are managed. Since road users do not pay directly for roads, they are not encouraged to choose whether and how to make a journey or, more particularly, to hold the road administration accountable for the way that it spends its budget. In addition, the absence of a firm link between revenue and expenditure encourages road users to demand more road spending *because* it is funded from general tax revenues and does not affect payments for road use. Without a hard budget constraint and pressure from road users, the road administration is not compelled directly to manage resources efficiently.

Concerns over road financing are leading many countries and governments to search for new funding instruments that address the above. Specifically, the issues being addressed are:

- Allocation of funds
- Alternative sources of revenue.

These are discussed in the following sections.

3.2 Allocation of funds

3.2.1 Budgetary process

Budgeting has two principal components:

- To decide how much money is needed
- To decide how to allocate the money that is actually awarded.

In most countries, budgets are allocated on an annual basis. A typical budgetary process might be as follows. There is an initial call from the Ministry of Finance for bids. Spending ministries respond to this, and the Ministry of Finance consolidates budget submissions from the various government departments, and reviews the submissions within overall spending targets. Heads of government ministries or departments may be called to support their submissions during this process. Draft estimates are then published, and are submitted to the elected government body for approval. Following this, a warrant for a given amount is issued to the relevant ministry or department so that spending may commence at the start of the new financial year. Often, the budget is released in tranches at fixed points in the year to prevent early exhaustion by spending departments.

Under such an arrangement, therefore, each ministry competes for funds and, at least in theory, funds are allocated to finance those expenditures with the highest economic or social return. However, such allocation processes are invariably highly politicised, and allocations are often far from economically optimal. Politicisation at national level is more likely to reflect general social welfare than at local level, where vested interests tend to have greater influence. Expenditures on maintenance, in all sectors, inevitably lose out to higher profile capital investment projects, which contribute to the under-funding of road maintenance noted earlier. As a result, road management budgets are often based on historical precedent: each year's budget is based on that of the previous year, with an additional allowance to cover inflation. This is a poor basis for budgeting, since it is arbitrary.

A better approach is for budget applications to be based on a rational assessment of economic need that relates to the objectives specified in the policy framework. One approach to needs-based budgeting is for budgets to be based on life cycle road costs. With this, an attempt is made to minimise the sum of maintenance costs, upgrading and reconstruction costs, and road user costs over the life of the road by choosing the optimum level of maintenance. If roads are maintained too soon, then the full value of the existing pavement will not be obtained and maintenance costs will be higher than any reduction in vehicle operating costs; hence, total costs are higher. Conversely, if roads are maintained too late, the consequent maintenance will be more expensive, or the value of the asset may be lost. Chapter 4 describes the concept of life cycle costs in more detail.

Thus the standards and intervention levels specified in the policy framework reflect the need for maintenance. If there is consistency between the policy for road maintenance and the budget, then the cost of the work needed to correct defects should be funded exactly by the budget. The data collection methods that can be used to determine maintenance needs are described in Chapter 5, and the methods for determining which maintenance treatments are the most appropriate are described in Chapter 6. In reality, the budget levels that are awarded seldom match the requirements of the policy framework, and priorities for expenditure must be set. Methods for this are described in Chapter 7.

There is often a wish by politicians to involve themselves in the detailed decision-making process of road maintenance. It is sometimes advantageous to turn this political interference into a positive force by cultivating the interest and inculcating a greater understanding of the maintenance process and its benefits. Where this has been done, increased budget levels have often resulted.

3.2.2 Budget heads

The financial provision for roads in many countries is divided between capital and recurrent expenditures. Capital provisions relate to the construction of new roads, and sometimes the reconstruction, rehabilitation, strengthening and resealing of existing roads. Recurrent revenue provision is for the regular maintenance of the existing road surfaces, off-carriageway features, and for dealing with various contingencies. Staff salaries are also normally paid from the recurrent budget. Strengthening and resealing works are more usually paid from the recurrent budget. Some road organisations have a more detailed breakdown of budget heads including, for example, specific budgets for the maintenance of bridges and structures, for surface dressing, and for 'environmental maintenance'. Where budgets that are awarded under different budget heads are less than those bid for by the road administration, the lack of ability to vire funds from one budget head to another may prevent the optimal allocation of resources under the overall budget constraint.

Road administrations are increasingly operating under a 'unified' budget provided by their paymasters. With these, there is no distinction between 'capital' and 'recurrent', or any other heading. A lump sum budget is awarded, and decisions on its expenditure between different types of works are made by the road administration itself on the basis of the policy framework. Local budget heads may then be used to assist in managing the funds. In general, a unified budget offers the best possibility for optimising expenditure, since this enables the needs of the network to be considered as a whole. It also enables funds to be vired between different budgets heads in the event of changing priorities during the year.

3.2.3 Allocation mechanisms

As noted in Chapter 2, roads of different hierarchies may be managed by different road administrations. Sometimes they will have their own sources of funds. For others, funding will come from national sources. Particularly for roads of lower hierarchy, the provision of funds will often be shared between national and local sources. It is common for central government to fund all work on national or trunk roads, but only part of the costs on other roads. In all cases, mechanisms need to be in place for allocating and disbursing funds between the different administrations. They need to be simple, transparent, and encourage consistency of standards between the different administrations. Three basic methods are commonly used for this (Heggie, 1995):

- Simple allocation formula
- Indirect assessment of needs
- Direct assessment of needs.

Simple allocation formula assign funds on the basis of pre-defined percentages to different parts of the network. For example, Japan allocates a fixed percentage of road-related revenues to different local government areas; in the past, Ghana spent 52 per cent of revenues on trunk roads, 28 per cent on rural roads, and 20 per cent on urban roads. Such allocation mechanisms are simple and transparent but, over time, are related only weakly to current need or use of the network.

Indirect assessment of needs is used where there are no reliable data for measuring need directly or where the cost of doing so would be disproportionate to the size of budget being allocated. The method is used mostly for the allocation of budget to low cost/low volume roads. Criteria used in the indirect assessment include:

- Land area covered by the administration
- Road density
- Population
- Agricultural production or potential.

The factors are weighted according to their perceived importance. This approach provides a pragmatic basis for allocating funds in appropriate cases that is cost-effective, and acknowledges the socio-political aspects of the decision-making.

Direct assessment of needs can be of different degrees of sophistication. At its most comprehensive, it will involve using the results of a detailed condition survey of all roads to determine works requirements. These are then costed to derive the budget requirement. This approach requires the use of treatment selection methods such as those discussed in Chapter 6. Simpler methods involve deriving norms for expenditure on roads in different hierarchies or of different surface types. Road lengths in each administration are simply multiplied by the relevant norms to give the budget allocation. Thus, there are several direct assessment methods. The method chosen should suit the capacity of the level of government concerned. The key is to have clear objectives, and then to appraise in a systematic way the extent to which each intervention contributes to realising these objectives.

It is often difficult to change existing allocation methods because there will be strong resistance from those who will lose out. There may also be pressures to maintain a 'regional balance' that may actually distort the optimal allocation of funds.

3.2.4 Disbursement mechanisms

Disbursement of budgets to road administrations can also be done in a number of ways, including the following:

- Funds disbursed directly to the road administrations
- Bills settled periodically after certification that approved work has been completed satisfactorily
- Funds disbursed on a conditional basis, with verification on the basis of technical and financial audits.

The above procedures have different requirements in terms of financial discipline and the degree of accountability. In this sense, financial discipline refers to the ability to enforce standards on those carrying out the work. It is a function of the incentives and punishment that are available for the control of quality, in the broadest sense. As a result, the different disbursement methods are likely to result in different qualities of the outputs produced by those organisations being funded.

The first method is the simplest, but has little financial discipline. Funds are simply disbursed directly to the road administrations who then have to account for the way in which they have spent the funds using the normal government accounting and audit procedures. Financial audit simply checks that funds received match funds spent, and the parent ministry is expected to ensure that the money has been spent on roads, and that the work has been done according to specification.

The second method requires more oversight of the way that funds are spent. It requires an approved work programme, together with technical and financial audits. It is normally used for work done under contract, and can also be used with in-house works. It involves considerable financial discipline, but is demanding in terms of works supervision and inspection. It is suitable mainly for major projects or works on the main road network.

The third method is suitable for a decentralised system of road administration. Funds are allocated directly to each administration, perhaps on a monthly basis, and their use is audited, typically at the end of the financial year by checking that requirements of quality and cost have been met. The results of the audit may then be used to help determine allocations of funds in subsequent years. This method also involves the requirement for significant financial discipline.

3.3 Revenue

3.3.1 Basic aims

The collection of revenue from road users has two main purposes (Newbery *et al.*, 1988):

- To charge road users for the costs that they impose both on the road administration, and on other users in terms of congestion.
- To raise revenues for government.

Effective financing for road management needs to satisfy a number of objectives:

- Provision of a secure source of funding to be used for road management, whose availability is certain and reliable.
- Independence from political interference on spending decisions, which should be based only on need, assessed using economic or other pre-defined criteria, and in accordance with the requirements of the mission statement.
- Establishment of a direct link between revenue contributions and spending on road management; with prices paid by users reflecting the level of service provided.
- Efficiency of revenue collection.

The concept of stakeholders and customers of the road network was introduced in Chapter 2. With ownership comes empowerment of road users and other stakeholders to take an interest in the management of roads. Support from stakeholders is a pre-condition for addressing the problem of road financing, since governments are usually reluctant to raise taxes and user charges (Madelin, 1996). Stakeholders in the network have their own vested interests. They may be willing to pay for roads, but only if the money is seen to be spent on maintaining and improving the network. In return, road users will expect the work to be done efficiently, and are likely to demand value-for-money. As such, involvement of stakeholders can create a surrogate market discipline to encourage the road administration to use resources efficiently and to discourage it from abusing monopoly power.

Ideally, such ownership can extend to become the basis of a genuine partnership, with stakeholders working with government on a number of issues. Examples exist where user organisations are collaborating with governments to improve road safety, control overloading, and to investigate innovative sources of funding (Heggie, 1995).

3.3.2 Principles of pricing and cost recovery

Objectives

Pricing and cost recovery policies should have four objectives:

- To use financing instruments that provide the correct market signals to road users
- To ensure that road administrations use resources efficiently
- To constrain the size and quality of the network to that which is affordable
- To generate sufficient revenues to operate and maintain the road network on a sustainable long-term basis.

Road tariff

In most countries, user charges contribute directly to general revenues, which are then used to fund work not only in the roads sub-sector but also in other parts of the economy. However, there are increasing pressures to move away from such a system (Heggie, 1995). An analysis of these issues suggests that roads would be managed better if considered to be part of the market economy, and if funds were secured from an independent source (Madelin, 1996). There needs to be a clear *price* for roads, and their management should be subject to market discipline. Road funding can be put on a similar basis to other utilities, such as telecommunications, electricity and water supply, where there is a direct link between users and suppliers. Such an approach encourages effectiveness and efficiency in the organisation and management of the supply, as noted in Chapter 2. Thus, road pricing should be considered as a 'utility charge'. Where road use charges contribute to general revenues, a defined portion of revenues, which relates to the cost of service provision, should be made available to the roads sub-sector.

Pricing and cost recovery policies are only likely to influence demand and to strengthen market discipline if the revenues collected from road users:

- Are seen to be spent on roads, and
- Impose a requirement on the road administration to meet defined and agreed levels of service.

To this end it is useful to consider road user charges as a *road tariff*, or utility charge, and not as a *tax*. This demarcates monies collected from road users into:

- A tariff that represents a use-related pricing mechanism, and
- Taxes that are paid into general revenues.

Special accounts, normally referred to as *road funds*, can be set up into which the proceeds from the road tariff can be deposited, and which can only be accessed by the roads administration. This prevents funds collected in this manner being spent on other public programmes. These are discussed further in section 3.4. It is important to note that putting roads on a fee-for-service basis, introducing a road tariff, and depositing the proceeds into a special account are not the same as conventional 'earmarking' (Johansen, 1989). As such, this approach addresses the macro-economic argument often made against dedicated, or 'earmarking' of, road funding.

It should be noted that Japan and the United States, both of which can be considered to have successful economies with well-developed budget systems, at least in part, secure road sub-sector funding in this manner to ensure adequacy and stability. In the United States, the trust fund has been in place for over half a century. It is broadly popular because it tends to reinforce notions of fairness, and concepts that the taxpayer at large should not be asked to pay for special benefits to certain groups (Johansen, 1989). However, in the United States a large proportion of road spending still comes from general revenues, and the trust fund has not protected the sub-sector totally from spending cuts. New Zealand is also considered as a good international example of fiscal probity, but also has a road fund.

A further advantage of the road tariff approach to funding has been found in those countries that suffer problems of late government budget approvals and no carry-over of funds. Improved synchronisation of funding resulting from the road tariff has enabled both better planning of works, and a more even cash-flow to contractors. This, in turn, has produced better competitive bidding and lower cost of works.

3.3.3 Sources of revenue

To maximise net economic benefits, it has been argued in the past that road user charges should be set equal to the cost of the resources consumed when using the road network, known as the 'short-run marginal cost'. This consists of the variable road maintenance cost, the costs of road congestion, and the external costs of environmental damage and road accidents (Heggie, 1991b). The problem with a short-run marginal costing approach is that only about half the costs of operating and maintaining a road network varies with traffic. The approach therefore requires that the balance of funds is derived from general revenues, unless congestion pricing is used to make up the deficit. It is now considered that the objectives listed earlier provide a better basis for pricing and cost recovery than the short-run marginal cost approach (Heggie, 1995). This essentially means that road user charges should be related directly to all costs of operating and maintaining the network.

The source of revenue that meets this requirement is a road toll. Levels of tolls can be set such that users pay directly for their contribution to the use of the road. This form of revenue collection is highly visible to users and gives clear market signals. Tolls can be levied by public administrations responsible for roads, and the issues involved are discussed in section 3.6. There is also an increasing interest in the private financing of roads, and tolls are also normally used in this situation to collect revenue. Private financing issues are discussed in section 3.7. However, the administrative cost of collecting tolls can be high and this method is not suitable for all roads. A relatively high traffic level is needed to make toll roads cost-effective, so other sources of revenue collection are more appropriate for the majority of the road network.

Further sources of revenue that can meet the above requirements are a fuel levy and a contribution from vehicle licences. A fuel levy reflects usage of road space directly and, when collected from the refinery or point of import, is an efficient funding source to administer. However, it is a poor proxy for road damage. Although heavy trucks use marginally more fuel than light vehicles, they do considerably more damage to the road pavement. If the main charging mechanism is a fuel levy, the contribution from heavy vehicles will be cross-subsidised by passenger cars. Some countries address this by charging a higher fuel tax on diesel fuel than on petrol, reflecting that most heavy vehicles are diesel-engined. A more direct reflection of the cost of pavement damage can be applied by using vehicle licence charges to provide a progressive fee that can match the relative degree of pavement damage caused by different classes of vehicle. International transit fees are also appropriate where such traffic is significant, and these can be collected at the point of entry and exit.

In urban areas, there may be a need for road pricing to reflect the cost of congestion. Mechanisms available for this include parking charges, 'cordon' charges and area licence fees. Electronic road pricing has also been tested in some countries. Congestion pricing is discussed in more detail in section 3.5. There is also scope for charging utilities for the use of road space, either when they are carried above the road, or buried within it. There is certainly a case for charging utilities when they are being installed or maintained, since the use of road space is being denied to the principal customers.

Many countries use a combination of the above financing instruments. The characteristics of the different charging mechanisms are given in Table 3.1.

3.3.4 Financing road maintenance and development

Road maintenance

The cost of maintaining different types of road varies considerably. This is illustrated in Figure 1.2 in Chapter 1. Charges based strictly on costs would involve wide differentials between different types of roads and the administrations responsible for their management. Although it is possible to maintain some differential between urban and rural areas and between different regions, in practice, some averaging will be necessary.

It is, therefore, necessary to set any fuel levies or licence fees carefully and fairly to address the above problems. The licence fee must be used to compensate for underpayments of fuel levy in terms of pavement damage. In other words, the licence fee cannot be used strictly as an 'access fee' to cover fixed costs. Instead the combined fee must be set carefully to meet both objectives. This requires careful analysis to ensure equity, and may need to change over time as relative prices and damage costs change.

Financing arrangements should focus attention on the affordability of a fully-funded road maintenance programme. Hence, it is usual to define a *core* network of strategic importance that is funded centrally. Non-core roads may either receive minimal maintenance, or possibly be handed over to lower levels of government, and funded from local sources.

Road development

Investments in extending the road network should normally be financed on the principle that whoever benefits should pay. Furthermore, increased road capacity to overcome congestion should, ideally, be financed from congestion charges. However, since this tends to be impractical, new investments must either be charged to all road users through the road tariff, or funded from general revenues through the development budget. Further thoughts on this issue are given in Box 3.1.

Table 3.1 Administrative characteristics of different road user charging mechanisms

Charging instruments	Potential role	Related to road use	Separable from general taxes	Easily recognisable	Administrative characteristics			Suitability[4]
					Collection cost (%)	Avoidance of evasion	Ease of collecting by contract	
Tolls	User fee	Yes	Yes	Excellent	10–20	Moderate	Simple	Poor
Vehicle licence fee	Vehicle access fee	No	Yes	Good	10–12	High	Moderate	Good
Heavy vehicle licence fee	Vehicle access fee	Not directly	Yes	Good	Unknown	Unknown	Simple	Good
Fuel levy	User fee	Partly	Can be	Good	Negligible	Low	Simple	Good
Weight–distance fee[1]	User fee	Yes	Yes	Excellent	5	Moderate	Moderate	Poor
International transit fee	Foreign use fee	Should be	Yes	Good	10	High	Simple	Good
Parking charges[2]	Control access	Partly	Yes	Good	>50	High	Simple	Poor
Cordon charge[3]	Congestion charge	Partly	Yes	Moderate	10–15	Unknown	Simple	Moderate
Area licence	Congestion charge	Partly	Yes	Moderate	10–15	Unknown	Simple	Moderate
Electronic road pricing	User or congestion charge	Can be	Yes	Good	<10	Unknown	Simple	Poor

Notes:
(1) A simpler form of weight–distance fee is the vehicle–kilometre fee; it employs the same basic principles, but relates fees more simply to vehicle type and distance.
(2) These are currently difficult to administer in developing and transitional economies and currently generate little revenue.
(3) These are only suitable when the road network lends itself to cordon pricing.
(4) This defines their suitability as *general* charging instruments.

Source: Heggie (1995).

> **Box 3.1** *Alternative considerations for funding new investment*
>
> If new investment is funded from the road tariff, there is a danger that this investment may take precedence over that for road maintenance, or that the road administration might undertake too much investment. Major new investment in the inter-urban network may have significant impacts on land use, location of industry and property values. This raises both strategic and political issues that should properly be dealt with by government.
>
> The counter-argument is that funding new investment through the road tariff forces users to pay the full cost of using the network, including the costs of investment. This should result in the size of the network being constrained to what is affordable, and will also allow essential investments to be carried out irrespective of the state of the government's budget.
>
> *Adapted from:* Heggie (1995).

3.3.5 Managing the revenues

It is important that any road tariff is administered efficiently. This means minimising evasion, avoidance and leakage; avoiding unnecessary subsidies; ensuring that any fuel levy does not tax inadvertently non-transport users of diesel; and minimising price distortions. Some comments on this are given in Box 3.2.

> **Box 3.2** *Notes on managing the revenues*
>
> Two main options exist to prevent road user charges and taxes from experiencing high levels of evasion, avoidance and leakage. The first is to simplify the tariff structure to reduce avoidance. In addition, licence fees can be focused on those for heavy vehicles. Since there are generally fewer of these in the traffic stream, and most are owned by registered businesses, such a fee is easier to administer. The second is to collect more fees under contract with the private sector, which introduces an incentive for collection efficiency.
>
> The tariff structure may also introduce inadvertent subsidies. For example, government and diplomatic vehicles often pay no fuel levies. However, these vehicles do impose measurable costs on the road network, and someone has to pay these costs. This creates distortions, and governments should recognise this by either ensuring that all users pay the appropriate fees, or by reimbursing the road administration for any lost revenues.
>
> Much diesel fuel is used outside the transport industry for power generation and to operate heavy equipment by industry. Such users should not be obliged to pay a levy on this that contributes towards a road tariff. In principle, differentiation between supply is possible; alternatively, rebates can be provided to non transport users of diesel fuel.
>
> If fuel levies raise prices, then this can encourage substitution between different types of transport fuels. Substitution can be discouraged by colouring of different fuels, but this is not entirely satisfactory. It is better to avoid wide price differentials between fuel types.
>
> *Adapted from:* Heggie (1995).

3.4 Road funds

Road funds are in use in a number of countries. They aim, in general, to influence demand and provide a basis for linking revenues and expenditures. The charges paid into a road fund should aim to be (Heggie, 1995):

- Related to road use
- Easily recognisable
- Easy to separate from indirect taxes and other service charges or fees
- Simple to administer, and not subject to widespread evasion, avoidance and leakage.

Box 3.3 *Pitfalls and problems of road funds*

The Ministry of Finance stops paying money into the road fund, holds up release of funds, or takes money out of the road fund and uses it for other purposes. This is generally caused by poorly designed arrangements for collecting road fund revenues, for paying them into a special account, or for authorising release. It may also be caused by a weak or non-existent road fund board, or ambiguous legislation.

If the road fund appears to be conventional earmarking of revenues, it is likely to be opposed by the Ministry of Finance. This generally happens when revenues deposited into the fund include general taxes and service fees. Such funds will be under constant threat of closure.

If insufficient revenues are available when the road fund has been set up, the result is that only part of the qualifying expenditure can be financed. The balance may then be financed through the government recurrent or development budget. However, once road administrations are in the position of receiving some money from a road fund, it often becomes difficult to obtain funds allocated through the normal budgetary process. Road funds should be set up to cover all qualifying expenditure.

Sometimes road funds can generate excessive revenues. This is rare, but can happen when the initial tariff has been set too high. Large surpluses are generated, encouraging other ministries to raid the fund or to bring pressure to have it closed. To alleviate this effect, arrangements are needed to vary the level of the road tariff.

There can be problems when the road fund lacks a firm legal basis, or is inflexible, and this can be a problem when legislation has been prepared quickly without adequate preparation. This can be avoided by establishing the road fund under existing legislation, and only passing new legislation when teething problems have been overcome. Alternatively, more time can be spent preparing the operational procedures before passing the legislation.

When there is no mechanism for revising the road tariff, other than through the normal tax-setting process of government, adjustments may require the approval of several ministries who have nothing to do with roads. This makes it difficult to adjust the tariff for inflation or to generate additional revenues.

There can be problems encountered when collecting the road tariff where revenues are collected by the Customs and Excise Department and channelled through the Ministry of Finance before being paid into the road fund. The best solution is to have the road tariff collected under contract and deposited directly into the road fund. Among other things, this emphasises its role as a user charge.

When the road fund finances roads managed by different administrations, transparent and equitable procedures are needed for dividing and allocating the revenues. Failure to do this results in allocations being erratic, or being subject to political whim.

There can be ineffective and inconsistent management when there is no road fund board, or when the board has inappropriate membership, or when it meets infrequently.

When the tariff is inconsistent it is not able to deliver a clear message to road users. Ideally, the road tariff should consist only of items that are clearly related to the road user, such as vehicle licence fees, the fuel levy, international transit fees, and bridge and ferry tolls.

Adapted from: Balcerac de Richecour and Heggie (1995).

Problems with road funds, where they do occur, tend to fall into two groups:

- Externally generated problems that affect the overall operation of the fund
- Internal problems that affect discrete parts of the operation.

These are elaborated in Box 3.3. Guidance on designing road funds is given in Box 3.4.

Box 3.4 *Designing a road fund*

Road fund management
The fund should be managed by a strong *Road Fund Board* with clear terms of reference; when one road administration is responsible for managing the entire road network, the road fund can be managed by the same board as the road network; otherwise there should be a separate board.

Contents of the road fund
The road tariff may typically consist of licence fees, a fuel levy, bridge and ferry tolls, and international transit fees.

Collecting the revenues
The road tariff should be collected by the Road Fund Board and deposited directly into the road fund without having to pass through the accounts of the Customs and Excise Department or the Ministry of Finance; oil companies can deposit the fuel levy directly into the road fund, and both licence and international transit fees can be collected under contract.

Setting the road tariff
There should be a formal mechanism for varying the road tariff, and charges should be indexed to ensure that they keep pace with inflation; the Board should either have the power to set the tariff, based on expert advice, or at least to recommend the tariff to the Ministry of Finance for inclusion in the annual budget statement.

Allocation of funds
There should be a simple and consistent procedure for allocating funds between different administrations entitled to draw upon the fund.

Auditing arrangements
The revenues handled by the road fund can be extremely large, so it is important to ensure that these sums of money are accounted for properly; independent financial and technical audits should be instigated to make sure that revenues are collected efficiently, with avoidance, evasion and leakage kept to a minimum, disbursed only according to approved expenditure programmes, and that work is carried out according to specification.

Adapted from: Heggie (1995).

3.5 Congestion pricing

3.5.1 Managing congestion

As traffic builds up and congestion occurs, the presence of each individual vehicle imposes a delay and, consequently, a cost on other vehicles. Urban congestion is characterised by regular peaked demands that are highest in the early morning and late afternoon periods. Increasing road capacity to cope with increased demand has proved to be largely ineffectual in urban areas: as road capacity has been increased, trips that were previously suppressed are released, filling up the additional capacity (*The Economist,* 1996). Ever-increasing demand for road space results in 'peak spill-over' and, in the worst cases, the 'peak' lasting all day.

There are three basic mechanisms for controlling urban congestion:

- Traffic restraint
- Indirect charging
- Direct charging.

All have been tried in different situations with varying degrees of success.

Any attempt to manage or control congestion should be undertaken within the scope of the policy framework, as discussed in Chapter 2, to ensure consistency with other sub-sector objectives. Lewis (1996) has suggested that aims in this area should be to:

- Act as a rationing mechanism for scarce road space
- Reduce non-essential travel demand and, consequently, the level of environmental pollution
- Provide a source of supplementary revenue
- Reduce demand for the construction of new roads
- Generate higher returns on privately financed infrastructure
- Level the field between private and public transportation.

A number of operational requirements should be considered when choosing a system for road pricing. These can be grouped according to the point of view of:

- The road users
- The road administration
- Society.

A summary of the issues to be considered is given in Box 3.5.

3.5.2 *Indirect charging mechanisms*

Congestion charges can be levied indirectly through vehicle ownership or use, or a combination of the two. Ownership charges have their place as a part of a general approach to road pricing, but are not very specific to congestion, and are not recommended for this purpose. Charging differential licence fees and fuel levies is difficult to administer, and is subject to abuse.

Parking charges are an appropriate and simple mechanism, but need to be supplemented with traffic management measures to prevent illegal parking and other types of avoidance. Parking charges may offer a transitional mechanism of congestion charging to ration scarce road space. It is perhaps ironic that users are often asked to pay a significant charge when the vehicle is parked and standing still, but pay no additional charge when the vehicle is moving and possibly contributing to congestion.

> **Box 3.5** *Criteria for choosing a road pricing system*
>
> *Road users' point of view*
> - Simple and convenient to use
> - Transparency with known pricing structure to enable motorists to make an informed choice
> - Legal and other safeguards to ensure anonymity of drivers
> - Options should be available for pre-payment, or payment in arrears through a billing system, allowing individual choice
>
> *Road administration's point of view*
> - Charges should relate as closely as possible to use, in terms either of distance or time
> - Prices should vary corresponding to rises in cost caused by changed demand
>   – temporally according to time of day, week or year
>   – spatially according to geographic location, route and, possibly, mode of travel
> - Operation should be reliable under harsh environmental or other adverse conditions
> - System should be secure against
>   – theft of funds
>   – fraud and abuse
> - Evasion should be difficult and enforcement easy
> - Pricing should be cost-based to reflect intensity of demand and revealed preference of travellers for certain times and for certain routes
> - Devices for registering transactions should be easily available, and provision should be made for infrequent visitors
> - A positive revenue–cost ratio is a necessary condition for commercial viability
>
> *Society's point of view*
> - The cost of implementation should be justified by economic cost–benefit analysis in which the opportunity cost of raising public funds or toll revenues has been taken fully into account
> - Road works necessary to install new systems should be kept to a minimum to avoid exacerbating existing congestion problems; similarly, environmental intrusion should be minimised
> - Prices should reflect congestion effects of different types of vehicles
> - Any new system should be phased in to soften its impact and to allow tuning to meet actual driver response
> - New systems need to be compatible with any already existing, such as
>   – public and private off-street parking
>   – origin and destination management information systems
>   – control systems of private firms' commercial vehicles
> - Pricing mechanisms are only likely to work where the population is law-abiding, with a culture of compliance to rules and regulations
> - Systems should be seen as fair by the public, and alternative means of transport available; this means that revenues must be ploughed back into transport through
>   – reduction in other road use charges
>   – better road system
>   – improved public transport
>
> *Adapted by:* H. R. Kerali *from:* Hau (1992a).

3.5.3 Direct charging mechanisms

Experience suggests that indirect costs, such as delay or environmental pollution, have only a limited impact on driver behaviour. The key to restricting the use of vehicles is to focus on the driver's perception of real cost (Lewis, 1996). It is, therefore, necessary to consider direct charging as the main mechanism for controlling congestion. This type of mechanism can be grouped into:

- *Off-vehicle recording*
  The charge is accumulated remotely, although there may be an in-vehicle electronic device triggering the charge.
- *On-vehicle metering*
  The charge is registered within the vehicle using automatic scanning and 'smart card' technology.

These have the purpose of charging on the basis of time or distance travelled. The following systems have been used (Ramjerdi, 1995; Lewis, 1996):

Off-vehicle recording using point pricing
- Manual charging by admission through toll gates and reserved lanes
- Automated scanning through automatic vehicle identification (AVI)
- Combination of the above.

On-vehicle metering
- Point pricing based on cordon
- Point pricing based on zone
- Continuous pricing by time and distance.

Note that the original cordon pricing system in Singapore was, in effect, a supplementary licensing mechanism, hence comes under the heading of 'indirect charging methods'. This has now been updated by the use of electronically operated entry points with in-vehicle smart cards for prepayment of congestion charges.

Technology in the area of direct charging methods is developing rapidly. Many innovative approaches, such as the use of 'fuzzy logic' (Hellendoorn, 1997), are also being investigated to address the intractable problem of urban congestion management.

3.6 Road tolls

3.6.1 Advantages

Road user charging systems that lack the power to discriminate between the location of road use, and can only imperfectly discriminate between different vehicles and forms of road use, can never achieve more than 'average' efficiency of road use. Some vehicles will be over-charged relative to the cost they inflict on the roads and other users; whereas others will be under-charged (Newbery *et al.*, 1988). Thus, in principle, tolling permits a much more accurate method of pricing than other systems of road user charges. Costs are charged to users specifically in relation to the use made of a particular facility, and charges can reflect use by different types of vehicle, carrying different loads, at different times of the day. Thus, each class of road user pays in relation to the costs they impose on the road network.

It is possible to set a toll to enable not only the recovery of all road costs, including financial cost, but also generate a revenue surplus that can be re-invested in the network. The extent to which this can happen will depend on the volume of traffic and the elasticity of demand which, in turn, is affected by the availability of alternative routes.

Tolling also increases the specificity of management because activities are focused onto a particular stretch of road that is subject to tolling. As noted in Chapter 2, increasing specificity is likely to increase the effectiveness and efficiency of management. This specificity may be increased further by involving private capital in the provision of roads, and this issue is discussed in section 3.7. The existence of toll roads alongside 'free' roads also enhances the specificity of toll roads. Potential users of the toll road will pay the toll in addition to all the other taxes and levies that are included in road user charges, but only so long as they consider that the improvement in service quality warrants the added expense.

Toll facilities may be administered either by public or private sector administrations and agencies.

3.6.2 Disadvantages

The presence of a toll road alongside those that are 'free' implies an element of double taxation. If tolls are imposed without any reduction in the real value of general road user charges, this is tantamount to imposing an additional general revenue taxation on transport. The higher effective level of taxation may be offset partly by lower vehicle operating costs if the road provides a higher level of service than the alternative route.

Where there are no 'free' substitutes for the toll road, an effective monopoly may exist that should be subject to regulation. A similar issue may arise if the alternative non-tolled roads are allowed to deteriorate, resulting in higher costs to road users. This will then strengthen artificially the position of the toll road operator.

However, it should be borne in mind that tolling is a relatively inefficient method of revenue collection from a public finance point of view, and can also have problems with leakage. The application of tolls may also result in cost increases to the road subsector. In any case, tolling is only likely to be feasible on a very small portion of the road network.

3.6.3 Pricing the toll

Economics dictate that road pricing should be directed towards maximising the use of the road (Johansen, 1989). To do this requires that the price, or user charge, should be related to:

- The fixed costs of using the network, which include the costs of traffic management and control, policing, enforcement, facilities management, and the provision of the basic administrative services that are not associated with road maintenance and development.

- The traffic-related road maintenance costs and the additional construction costs required for heavy vehicles.
- Any costs imposed on other users if the additional trips generate congestion due to capacity constraints.

When toll roads become congested with increased traffic levels, congestion pricing becomes appropriate (Hau, 1992b). This should normally be geared to raising revenue to provide additional capacity to relieve the congestion. Tolls should therefore be set at a level to achieve this, and should reflect the reduction in the level of service on the existing road. However, in this situation, tolls are normally used as a means of rationing limited road space.

Other pricing criteria contribute to efficient operations from both an economic and a financial point of view. These include:

- Toll rates should be set specifically for each toll facility and linked to the specific traffic, operating conditions and construction costs; this may involve having variable tolls on an individual facility to reflect, for example, peak and off-peak use.
- The need for toll rates to be adjusted periodically to reflect changes in overall prices; an agreed inflation index is normally adopted for this purpose.
- When service levels reduce as a result of congestion, toll charges should be increased to limit congestion.

3.6.4 Investment criteria

The basis of the economic investment decision for toll roads is no different from that for any other type of road. These criteria are discussed in Chapter 4, and apply whether the construction of new roads is being considered, or whether investment in tolling is to be introduced to fund the maintenance and operation of an existing road. However, there may be economic losses, as a result of tolling a road, that need to be taken into account in the appraisal. These include:

- Additional investment cost to pay for added lanes for toll booths, the booths themselves and any toll collection equipment.
- The additional cost of toll collection and administration compared with the mechanisms of financing a conventional road.
- The extra costs imposed on road users as a result of slowing down or stopping to pay tolls; this will include increased vehicle operating costs and time costs due to the extra speed change cycles.
- Additional environmental costs in terms of noise and air quality because of vehicles slowing down and accelerating.
- The additional costs imposed on road users who are diverted to alternative routes when an existing route is re-designated as a toll road.

The extent of these will depend on the type of system introduced, its operating conditions and prices. Furthermore, they are interdependent, as illustrated in Table 3.2.

Table 3.2 *Interdependence between impacts of tolling*

System elements	Affected costs	Design and management strategy
Toll system and design	• Investment • Toll collection • Road user costs	• Minimise investment • Minimise collection costs (perhaps by using an open system) • Minimise use of staff to lower operating costs
Operating	• Toll collection • Road user costs	
Toll rates	• Diversion • Non-toll road users	• Set rates on an elasticity basis • Monitor performance

Adapted from: Revis, J. S. Toll road pricing, financing, investment and design criteria for developing countries (in: Johansen, 1989).

The provision of a toll facility, providing a higher standard than those that already exist, is likely to result in traffic diversion. Such diversion is also likely to occur as a consequence of change in prices or general economic conditions. It will affect not just the toll revenue, but also the wear and tear on other roads. This needs to be taken into account in the social cost–benefit analysis of a potential scheme.

However, it is worth noting that the decision to impose tolls on a road is not a commitment to a particular way of financing its construction. A variety of financing options may be available, each with advantages and disadvantages in terms of public finance, economics, administrative efficiency and political feasibility. Private financing options are discussed in section 3.7.

3.6.5 Design criteria

Three main options are available for toll collection (Johansen, 1989) which affect the cost of toll operation and the extent to which the road can be accessed by users:

- *Closed system*
  Tolls are paid on the basis of trip length by collecting a ticket at a barrier on entering the road, and paying at a booth on leaving the road.
- *Open system*
  Toll booths are placed at strategic locations along the road at which users pay their tolls as they pass.
- *Mixed system*
  The road contains both open and closed sections, which depend on terrain, traffic and congestion conditions.

The closed system restricts entry and exit to a relatively few locations because of the need to collect tolls at each access. This system tends to divert all short distance trips (less than 15–20 km) to local roads, as well as significant numbers of medium distance trips (20–30 km) because the limited number of access points are not always convenient for origins and destinations of trips. Thus, the number, location and distance between access points will determine the extent to which the road is used, and by whom.

For the open system, since tolls are collected at booths placed along the road, the number of accesses can be the same as for an ordinary road of a similar standard. Some short and medium distance trips can take place free of charge. It is often convenient to place toll booths near to the centre of urban areas through, or around which, the toll road passes, with no interchanges in these localities. Such a design protects the road from encroachment of traffic from the urban areas. Alternatively, toll booths can be placed in between the boundaries of urban areas, with accesses within the urban areas. This design encourages local traffic, but may cause congestion as traffic levels grow.

A well-planned open system has toll booths at points where the largest population of vehicles are travelling long distances. These vehicles receive greatest benefit from the road, and this design approach reduces diversion to other routes. Experience suggests that closed systems cause more diversions and, as a result, have a more depressing effect on local economies than open systems.

Studies indicate that collection costs on closed toll systems can be as high as 15 to 20 per cent of gross revenues (Johansen, 1989). On open systems, costs are substantially lower, typically under 5 per cent of revenues. Toll authorities have introduced a number of management methods to reduce collection costs, including:

- Greater automation of toll collection and a flat rate toll
- One-way collection on bridges
- No collection during off-peak periods
- Use of electronic billing.

3.7 Private financing

3.7.1 Background

In many countries, there has been considerable interest in recent years in the private finance of road infrastructure. This interest has been driven by the need for governments to finance infrastructure modernisation off-budget, and to obtain inputs both to a large pool of additional funds and private sector management skills. The interest has arisen both in regions experiencing rapid economic change and in mature Western economies. The approach has stimulated most interest where there is seen to be the need for roads with both high quality and capacity. Private financing includes private concessions, public toll roads, private–public partnerships, and community fund raising by targeting local taxation (Farrell, 1994).

Private finance has been seen by some countries as a panacea for addressing problems of limited government finances. The interest has been fuelled by contractors who have been short of work during times of economic recession. However, several issues are raised by experiences of initiatives so far taken. Some of these are given in Box 3.6. As implied above, private finance initiatives are only ever likely to be viable on roads carrying high traffic volumes. Even so, the initial investment required is high, and a long time period is needed for cost recovery. In countries with uncertain political and economic situations, contractors, banks and private financiers are unlikely to be prepared to risk the considerable sums of venture capital needed to finance such activities. Nevertheless, there are many cases where private financing has proved to be successful.

Normally issues of private financing in the roads sub-sector relate to 'concessions'. A concession is the award of a right or a licence to build, own and operate a public service for a given period (Farrell, 1994). In financial terms, 'concession financing' is the design, arrangement and implementation of a financial package for a major

---

Box 3.6   *Experience of private finance initiatives*

Over-optimism
There has been a tendency, at least for motorway construction, to launch over-ambitious concessions, and then to have to scale them down because of high costs of the projects and the overheads.

Project staging
A conventional *build, operate and transfer* (BOT) project precludes major project staging, and may therefore require uneconomic premature investment.

Innovation
There has been little evidence of private sector innovation in BOT projects, possibly because rules set up for their control have been too constrained.

Equity
'Good projects', in financial terms, have attracted investors who are able to protect the interests of the concession company. However, 'poor projects' have tended to attract only contractors, and this has created conflicts of interest.

Project costs and revenues
The psychology and mechanics of concession negotiations with contractor-sponsors provides opportunities to increase prices. There is little incentive for the client to negotiate prices down, provided that banks are prepared to finance deals. There has been a tendency, in the event of a funding-gap, to 'adjust' the revenue forecasts upwards to make the project bankable. Ultimately, banks have to decide whether they believe the revenue forecasts and, therefore, whether the debt service coverage rates are likely to be met.

Economic analysis
Appraisal of economic consequences plays no material role in assessing BOT concessions provided that the BOT is essentially 'cost-free' to the state and that there are no subsidies. However, when state subsidies are involved, investment policy issues arise and economic appraisal is then appropriate. Issues then arise about whether investment decisions should be aiming to maximise financial or economic returns, and these can be in conflict.

Complexity
BOT deals are highly complex and time-consuming. This makes them expensive to implement, and consultancy and legal fees can easily exceed one million dollars.

Adapted from: Smith G. (1995).

project, in which the developers will consider the future cash-flow of the project and its assets as the principal forms of security. Sometimes financial guarantees are necessary from interested third parties, often governments, to cover certain perceived risks. In the event that the credit structure is free from guarantees from third parties, it is termed 'non-recourse'; where partial financial guarantees are available, the term 'limited recourse' is used.

The term *BOT* stands for 'build, own and transfer' or 'build, operate and transfer'. Other terms in common use include *BOOT*, 'build, own, operate and transfer'; *BOO*, 'build, own and operate'; *BOOST*, 'build, own, operate, subsidise and transfer'; *BLT*, 'build, lease and transfer; and *DBFO*, 'design, build, finance and operate'. All are types of concessions.

3.7.2 Advantages

Several advantages are often cited for the use of private finance in the roads sub-sector.

Projects tend to be brought on stream more quickly than they would be in the public sector because of constraints on government spending.

The cost is borne by those who use or benefit from the facility, not by the taxpayer.

Improved efficiency during construction and operation may result, along with the introduction of new sources of skills for constructing, operating and maintaining the facilities as a result of private sector involvement. However, there is evidence that use of the private sector is no more efficient than the public sector, particularly when operating through a toll road authority or agency.

Access may be provided to new sources of funding. Private concession agreements for toll road financing may facilitate the raising of substantial amounts of private capital and, where foreign investment is participating, this may result in a larger aggregate investment in roads than would be possible from purely government sources. Where equity financing is used, investment may be possible at cost levels that reflect different operational criteria and financial mechanisms not normally available to the public sector.

There is usually an improved recognition of project risk by governments, because the negotiation of concessions forces governments to face up to issues that would rarely be addressed when preparing public sector funded projects.

The concession company is devoted to developing its own cadre of employees who have a vested interest in the commercial development and operation of the concession.

Privately financed schemes are felt to generate new ideas and concepts because of the need for innovative approaches to all aspects of project preparation and operation. In addition, private financing is considered to be more flexible, and to provide the opportunity for experimenting with new operating methods.

3.7.3 Disadvantages

However, there are also some disadvantages associated with private financing.

Road infrastructure has certain characteristics that are unattractive from a private sector perspective. Investments are often high cost, high risk and not easily divisible into smaller, more manageable units. They also frequently have long gestation periods for both planning and construction, which tie up working capital.

Concession agreements using private financing may be more expensive in the short term because of the returns expected by private bankers operating internationally. Therefore, financing charges are generally higher than those in the public sector. However, for longer-term investments, expectations may not be very different from those in the public sector.

Provision of infrastructure under concession arrangements is more complex than with conventional methods. Among other things, this can result in the time taken for implementation being longer than publicly-funded works.

There may be conflicts between the private developer and government because of their different objectives. The financier will be seeking to maximise the financial return on the investment, while the government will seek to maximise the social welfare and economic benefits generated by the project. Unless strong conditionality is built into the agreement, the concession process can sometimes be used as a means of by-passing the normal public budget process and avoiding the application of economic efficiency criteria for investment and pricing policies. The budget process can be an effective mechanism for preventing the mis-allocation of funds to low priority projects. To avoid the concessionaire trying to maximise financial return, rather than meet social development objectives, the government may need to establish clear public service level guidelines in the agreement and, conversely, may have to compensate the concessionaire for loss of revenue.

Often concession agreements require government guarantees against risks, such as:

- Traffic volumes being less than forecast, leading to shortfalls in revenues
- Government intervention through policy changes that reduce revenues or increase costs
- Exchange rate fluctuations or devaluations that could increase costs to concessionaires
- Inflation, which typically increases costs faster than revenue.

Without such guarantees, financial rates of return may need to be high in order to attract private financing. Sometimes the government may take on some of the risk itself, rather than provide guarantees, thus entering into a partnership with the concessionaires.

3.7.4 The parties involved

Government and the private sector each have distinctive roles under private financing initiatives (Hamilton, 1996). The role of government can be summarised as follows:

- Ownership of the road; the government has an interest in ensuring that its assets are well maintained.
- Safeguarding the public interest through regulating the use of monopoly power and by maintaining safety standards.

- Indirect financial support through a combination of subsidies, tax concessions and direct equity investment.
- Authority to provide competing and complementary roads
  – competing roads can harm the concession
  – complementary roads, such as a feeder road, can improve the concession's profitability.

The role of the private sector concession company is to defray the up-front capital cost away from government, and to provide a financial return to shareholders. The main private sector participants are:

- Shareholders in the concession company, often including construction companies.
- An operating company which manages and maintains the road to the required standard, and ensures that the toll collection facilities operate at an agreed performance level; this company may be part of, or separate from, the concessionaire.
- Banks and brokers seeking to provide high-return, non-recourse debt finance.
- Companies involved in road works seeking to undertake large contracts which otherwise they might not win.

Both government and private sector institutions must accept the risks associated with this type of finance. These are summarised in the following sub-section.

3.7.5 Contractual framework and risks

Developing a privately financed project is a complex task because of the many parties involved. The mix of financial and contractual arrangements needs careful structuring and expert negotiation (Keong *et al.*, 1997). With private finance initiatives, the promoter takes on a much greater role than in traditional contracts. To be effective, the framework for formulating and awarding a contract or concession should be (Farrell, 1994):

- *Realistic*
  Governments must be aware that private investors expect to make a profit, and will not take risks unless the returns are very high; the private sector must accept that governments are bureaucratic, and need to follow procedures that are designed to protect the public.
- *Rigorous*
  The project brief, prepared by government, needs to establish clearly the shape of the project, its regulatory environment, and the selection rules; rigorous project preparation should be translated ultimately into a better deal for the public.
- *Competitive*
  The role of competition for effective and efficient procurement was noted in Chapter 2; schemes for private financing will also benefit from this same approach.
- *Transparent*
  Many deals are negotiated, opening up opportunities for fraud and corruption; the negotiated element should be kept to a minimum so that the award process can be as open and transparent as possible (see also Chapter 9).

The issue of risk also needs to be considered (Hamilton, 1996). The first concern is that the private sector may not be able to recover the costs spent on project preparation. These costs may be significant, with high front-end loading and, at best, there is likely to be a long wait before the investment is paid off. Since investors are committing considerable sums to the development of an idea, they may consider that they have a 'right' to proceed with it on an exclusive basis, or at least be reimbursed for the initial expenditure. There is also the risk of higher-than-expected construction or financing costs, or that government may impose a relatively low cap on allowable levels of profit.

Road financing also has specific risks. These include the absence of an alternative use for a road if demand proves to be insufficient, and the low value of a partially constructed road if there are pressures on finance which prevent or delay completion. Once money has been sunk into project preparation, the developers may wish to make rapid progress with implementation; delays cost money and increase risk. Particular problems relate to both the procurement of land and planning procedures, which can be difficult, cumbersome and slow.

There are also civil engineering risks. In privately-financed schemes, these are all borne by the developer. This is unlike public sector schemes where risk is shared on a contractual basis between client and contractor.

However, the greatest risk is in cash-flow, with the possibility that revenue over the life of the concession will be lower than expected. Demand forecasting is still an inexact science, but demand is linked to economic growth, which is particularly difficult to predict. Demand may also be related to other developments in or outside the sector over which the developer has no control.

Risk can perhaps be minimised by:

- Using analysis procedures based on risk assessment (see Chapter 1)
- Considering syndicated data collection programmes
- Being prepared to back away from projects where the risks are too high relative to the rewards, or where the associated terms are unattractive.

For successful private financing, a key step is the development of an appropriate financial package and the raising of the necessary funds. Several criteria can be used to help judge the suitability of projects for private financing, and some of these are given in Box 3.7.

3.7.6 Shadow tolls

Initiatives have been taken in the United Kingdom for private sector financing of roads with costs being recouped by the developer, over time, using 'shadow tolls'. Financing follows conventional methods of private sector concessions, as discussed above. The innovation is in the way that the tolling mechanism is set up. Instead of physical tolls being collected, this is replaced by a shadow toll. This is an annual fee paid by government to the concessionaire for managing the road. The size of this fee depends on the level of traffic actually using the road during the year.

The concessionaire has the responsibility for managing the road including all maintenance works. The purpose of this approach is to move the capital cost of such

> **Box 3.7** *Conditions for success of private finance initiatives*
>
> *Economic stability*
> - The country where the concession is located must have a strong and stable economy
>
> *Financial viability*
> - There must be adequate financial returns, with the ability of the cash-flows to meet debt service and investor returns, plus a margin for safety
> - The foreign exchange implications
>
> *Attitudes towards risks and rewards*
> - The risk that the project will not be completed according to its original cost estimates and construction schedule
> - The risk that the project, after completion, will not be maintained adequately or operated efficiently
> - Financial and political risks
>
> *Completion undertakings*
> - The need for comprehensive completion undertakings, including the use of performance bonds
>
> *Availability of finance*
> - Local financial markets must be well developed, and there should be support for the concession from local financial institutions
> - The availability and amount of equity must be assured, including sufficient to meet interest payments
> - Debt will normally need to be syndicated to spread risk
>
> *Government support*
> - There must be political will to succeed
> - Creation of a legislative, regulatory and taxation environment within which concessions can operate effectively, and which balances risks and rewards
> - Providing adequate and qualified personnel to overcome bureaucratic delay, and to expedite legislative issues, while representing the public interests at negotiations; and thereafter to minimise government interference once the project is under way
> - Providing clear and concise public participation procedures in the planning process, and by dealing with environmental issues
> - Providing some protection for a limited period from the possibility of competing facilities, and providing some protection against *force majeure*
>
> *Complexity*
> - Concessions can be structurally complex and require extensive documentation, with the result that relatively small projects become too costly and time-consuming to implement
>
> *Patience and perseverance*
> - Concessions cannot be implemented quickly and, whereas governments may be prepared to spend time, private developers are often constrained to produce results in a shorter time frame
>
> *Adapted from:* Blaicklock, T. M. Financing infrastructure projects as concessions (in: Farrell, 1994), and Keong *et al.* (1997).

works from government to the private sector. The level of fees negotiated aims to cover the maintenance costs, plus repayment of the initial investment, including interest, over the period of the concession which is relatively long. The higher the traffic, the higher the management fee paid to the concessionaire, although the details of the concession agreements vary.

The UK Government's declared aim with this approach is to 'foster the development of a private sector road operating industry' (Hamilton, 1996). Although the approach represents an interesting way of reducing demands on a government's capital budget, it must be recognised that the approach does not raise extra revenue. Essentially, the approach commits, or earmarks, funds allocated for roads into the future. It can be considered as a long-term maintenance contract, with payments indexed in terms of traffic, compared with a normal 3–5 year term contract (see Chapter 9).

3.8 Local financing

Roads that have not been 'adopted' or 'proclaimed' by the road administration are considered to be private roads and, in general, receive no public support for their construction or maintenance. These roads might be used by logging companies, commercial farmers, or local areas of housing. Formalising the management of these by adoption by the local interest group often requires the following:

Box 3.8   *Private co-operative roads in Finland*

Finland has over 100 000 km of private roads that are managed by co-operatives. These roads, mostly with gravel or earth surfacings, carry an average of about 45 vehicles/day. The *Private Roads Act* sets up a mechanism for the management and funding of these roads by the users and residents themselves, with support from government. The Act stipulates the right-of-way, the co-operative ownership, and the formula for distributing maintenance costs among both road users and adjoining property owners. The co-operative is responsible for organising maintenance of these roads, and may either pay its own members to do the work, or may use a contractor. Membership of the co-operative is compulsory for property owners who use the road.

The co-operative sets its own maintenance fees, accepts new members, and is responsible for having the previous year's accounts audited. Maintenance costs are shared among members, depending on the size of their property and the amount of traffic that they generate. The government supports maintenance of co-operative roads provided:

- A formal co-operative has been established
- The road length to a permanent residence is at least one kilometre
- There are at least three estates with permanent residents along the road

In 1990, government support to co-operatives was $30 million, support from municipalities was $40 million, while the amount paid by the co-operatives themselves was $50 million. Government support is channelled through the Finnish National Road Administration, and allocations are based on traffic volume and the number of permanent households serviced. The figure is also adjusted for climate and average income.

Adapted from: Korte (1995).

- A legal framework that enables governments to cost-share
- An incentive system to persuade the individuals or groups to adopt selected roads
- Access to advice and technical assistance to help them carry out the management function efficiently
- Oversight arrangements for technical and financial supervision to provide accountability for the use of any public resources.

A 'private streetworks' act may need to be passed before governments can enter into the required cost-sharing arrangements. Without this, it may be difficult to put in place appropriate funding arrangements.

The incentives usually take the form of cost-sharing arrangements. The road administration pays part of the costs of maintenance, and the adoptive owners of the road agree to pay the balance. To participate in such an arrangement, the owners need to enter into an agreement with the road administration and agree to meet any requirements that are laid down. Owners may need advice, technical assistance and training in planning, contracts and works execution, and advice on how to deal with unexpected problems that might arise during implementation. There is also a need to ensure that works are carried out to an agreed standard and that resources provided are accounted for properly. Technical supervision is normally provided by the road administration, and accounts are normally required which must be audited.

There are several examples of countries that have introduced procedures that encourage the owners of unadopted roads to register their interest in a specific road in return for support for its maintenance. In Finland and Sweden, adjoining land owners are encouraged to form themselves into road co-operatives for managing private roads. In Ontario, Canada, local roads boards are encouraged and individuals are invited to register as owners of roads. In Namibia and South Africa, there are special arrangements under which commercial farmers receive assistance to enable them to maintain unproclaimed roads, provided such roads serve more than one landowner. The examples of Finland and Sweden are discussed further in Boxes 3.8 and 3.9.

> **Box 3.9** *Managing the private road network in Sweden*
>
> Over half of the Swedish road network consists of private roads. Of these, 74 000 km receive State subsidies, and 210 000 km are not subsidised. Those that have been granted State subsidies are the responsibility of 24 000 road managers. Fifteen road managers responsible for ferry routes also receive State subsidies. Most of the unsubsidised private roads are forest roads.
>
> There are approximately 330 000 dwellings and 800 000 people living alongside roads that are eligible for subsidies, excluding those located on side roads. In addition, slightly more than 220 000 recreational homes are situated on the subsidised private road network.
>
> Approximately 60 000 km of the subsidised private network are used daily by the general public for commercial as well as leisure activities. The private road network generates a large proportion of private and goods traffic, with almost one in every four vehicles having its origin on State subsidised private roads. Short trips mean that total vehicle-kilometres is lower on the private as opposed to the public road network. Nonetheless, about one-third of the private road network has a traffic in excess of 100 vehicles/day. In addition, approximately two million trips are made each day from permanent and recreational homes. This translates into 1 700 000 vehicle-kilometres in a 24 hour period, to which can be added approximately 700 000 trips by 'unknown' traffic. Statistics relating to the private road network are summarised below.
>
> | | Length of road in each category (km) | Maximum State subsidy (%) | Annual costs of upkeep (SEK million) |
> |---|---|---|---|
> | Roads giving access to permanent residences in rural areas | 53 000 | 70 | 375 |
> | Roads giving access to permanent residences in urban areas | 2 000 | 40 | 7 |
> | Through roads | 7 000 | 80 | 55 |
> | Recreational roads | 3 000 | 50 | 11 |
> | Roads giving access to recreational properties | 3 000 | 40 | 15 |
> | Roads giving access for commercial activities | 5 000 | 70 | 37 |
>
> *Note*: US$1 = approx. SEK18.
>
> Tax revenues on fuel consumed by traffic on the private road network amounts to approximately SEK850 million (US$105 million) per year.
>
> *Source*: Swedish National Road Administration.

4 Benefits and Costs

- Basis for management decisions
- Benefits
  - impacts and beneficiaries
  - project and network level benefits
- Costs
  - methods of accounting
  - cost estimating techniques: global, unit rate and operational costing
  - suitability of individual techniques
- Cost–benefit analysis
  - prices
  - comparison of alternatives
- Life cycle costing
  - road administration costs
  - road user costs
  - life cycle cost models

4.1 Basis for management decisions

While a number of factors will have to be taken into account when taking management decisions in the road sub-sector, a key factor will always be that of value for money. Decisions will need to be taken to ensure that limited funds are spent to greatest effect within the various constraints that pertain. For many decisions, there will be a need to determine the costs of implementing the particular decision, and to estimate the benefits resulting from this.

Wherever possible, benefits and costs should be quantified so that there is confidence in the basis for the decisions. Money is possibly the only mechanism available for comparing the value of dissimilar items. It is less abstract than concepts such as beauty and comfort. However, even for physical items, the use of money as an indicator depends on there being a free market with willing buyers and sellers. Where it is possible to quantify costs, then any savings in these costs represent benefits.

Whereas the costs of undertaking specific activities are relatively easily identified, predicting the benefits resulting from expenditures is often difficult. As a result, there is only a limited amount that can be written here about benefits, and the balance of this chapter discusses issues surrounding costs. There is a need for studies in the areas of identifying and quantifying benefits from improved road management. Nevertheless, these issues are discussed in the following sections.

4.2 Benefits

4.2.1 Impacts and beneficiaries

Benefits from better road management can be expected to accrue in those areas, identified in Chapter 1, where impacts would result from works on the road network:

- Level of service or road condition
  - road surface conditions and general driving environment that are more appropriate to the road function
- National development and socio-economic impacts
  - economic growth and improved national well-being
- Road user costs
  - reduction in vehicle operating costs and savings in travel time
- Accident levels and costs
  - reduced numbers of accidents, savings in death and injury, and reduction in consequential costs
- Environmental degradation
  - a healthier, less damaged, and more pleasant environment for those using and living adjacent to the road
- Road administration costs
  - better value for money and lower costs for activities undertaken on the road network.

The beneficiaries of impacts from better road management will be the stakeholders identified in Chapter 2:

- Owners and operators of commercial vehicles and buses
- Representatives of industry, commerce and agriculture, who have a vested interest in an efficient road network to support their business operations
- The travelling public using the road network
- The road administration and the road engineering industry.

The actual size of benefits obtained under the headings, above, will depend on the particular policy framework of the road administration, and on many other factors. For example, benefits from road management are likely to vary from region to region, depending on the particular topographic, climatic and socio-economic situation in the region (Mosheni *et al.*, 1994). Benefits will vary on roads with different traffic levels: roads with higher flows are likely to produce higher benefits per unit of road maintenance expenditure than roads with low flows.

It could be considered that, in general, the potential size of benefits will relate to the levels of expenditure in different areas. Figure 1.1 showed typical relative sizes of cost for road maintenance, vehicle operation, travel time, accidents and the environment. However, in addition to these potential direct savings in cost, it has been noted that although expenditure on road maintenance is relatively low compared with other road costs, improvements in road maintenance can result in considerable cost savings in all areas where costs are incurred.

The nature of benefits differs, to some extent, depending on whether decisions are being made in connection with planning and programming ('network level'), or with preparation and operations ('project level'). It is, therefore, appropriate to consider benefits under these two headings. The 'project level' case is more straightforward, so is considered first.

4.2.2 Project level benefits

Depending on the particular road management activity, it is possible to be more specific about the benefits likely to be obtained, when considering activities at project level.

For example, the benefits arising from repaving a road could be grouped into the following categories (TRRL Overseas Unit, 1988a):

- Lower vehicle operating costs because vehicles will be driving on smoother running surfaces
- Road maintenance benefits because fewer maintenance inputs will be needed on the new pavement surface
- Savings in travel time because vehicles will be able to travel at higher speeds
- Reduction in road accidents because, for example, motorists will not need to weave or to change lanes to avoid pot-holes.

Other examples of benefits obtained at project level are discussed elsewhere in this book, including benefits from:

- Timely road maintenance and improvements to road riding quality (Chapter 1)
- Axle load control (Chapter 1)
- Collecting data at an appropriate 'information quality level' (Chapter 5)
- Choosing optimum grading frequencies for gravel roads (Chapter 6)
- Using a decision-based approach to contract packaging of works (Chapter 6)
- Using works prioritisation methods that take into account the future implications of decisions (Chapter 7)
- Planning routine maintenance requirements on the basis of inventory records (Chapter 8)
- Rationalising snow clearing and de-icing of routes, and utilising timely and appropriate weather information for winter maintenance (Chapter 8)
- Using competitive procurement for both road maintenance works and their management (Chapter 9).

The quantification of benefits at project level is normally relatively easy.

4.2.3 Network level benefits

Whereas project level benefits tend to be relatively discrete and definable, those at network level are usually less tangible. Yet it is at network level, when dealing with the functions of planning and programming, that consideration of benefits has a key role to play in helping to frame policy and to argue for funds. Network level benefits also derive from institutional issues, such as those from:

- Increased effectiveness and efficiency through commercialisation (Chapter 2)
- Increased and more effective outputs through training (Chapter 2)
- Use of improved road financing mechanisms (Chapter 3).

There will be some cases when it is possible to aggregate benefits from a number of discrete projects to derive a figure that applies to the network as a whole. However, the individual and discrete nature of projects usually means that, by definition, it is difficult to extrapolate their benefits across all sections in a wider network. Generally, different approaches are necessary.

One of the biggest problems is how to present network level benefits in a simple and straightforward way that is meaningful to politicians and senior decision makers. Modern network planning tools, for example, 'NETCOM' (Kerali and Snaith, 1992), provide a convenient means of doing this. These tools can produce predictions of items, such as future network condition for a given budget level (Figure 4.1), and the budget required to maintain a constant network condition (Figure 4.2). Producing such graphs for different budget and condition scenarios can be used as the basis of a dialogue with political decision-makers on the network level benefits of adopting certain policies.

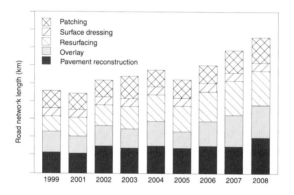

Figure 4.1 *Future network condition for a given budget level* (*Based on*: Kerali and Snaith, 1992. Crown copyright 1992. Reproduced by permission of the Controller of H. M. Stationery Office)

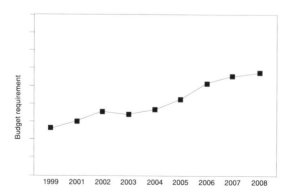

Figure 4.2 *Budget required to maintain a constant network condition* (*Based on*: Kerali and Snaith, 1992. Crown copyright 1992. Reproduced by permission of the Controller of H. M. Stationery Office)

A further example is given in Figure 4.3, which shows target and actual levels of condition for different percentages of the road network. Where such targets are based on minimising the sum of road administration and road user costs, any deviation from the target levels represents an economic loss to society. It would be expected that different targets would be optimal for roads of different hierarchy.

For effective road management, every effort should be made to identify and quantify the benefits associated with each activity undertaken to enable its worth to be assessed.

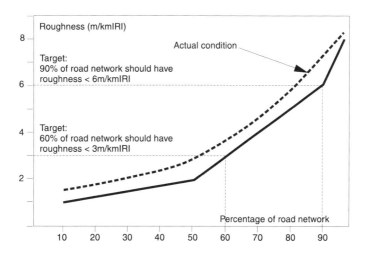

Figure 4.3 *Example of target condition levels for road network*

4.3 Costs

4.3.1 Issues to be considered

Determination of costs needs to take into account several factors. These include:
- The method of accounting used
- Consideration of the purpose for which the costs or cost estimates are used depending on the management function
- The accuracy to which costs, and particularly cost estimates, are needed.

Determining costs that have already been incurred is a matter of keeping records in an appropriate format, and of having operational procedures for recording costs in a prescribed manner. These records then provide a data source that can be analysed to determine expenditures. These may be broken down under headings and sub-headings that facilitate management decisions and control. However, accounts need to be kept in a form that enables costs to be assigned to activities undertaken.

Such records also provide a useful source of information when predicting estimates of likely future expenditures. However, estimates of different levels of accuracy may need to be produced for different requirements. The purpose for which costs or cost estimates are needed should determine the manner in which costs are collected and estimates made. A particular problem for many road managers working within the public sector is the requirement to use methods of accounting laid down by ministries of finance that are not always the most appropriate for local management purposes.

4.3.2 Methods of accounting

Much road management work is undertaken in the public sector where costs are allocated according to systems of government accounting that are stipulated by ministries of finance. Such systems are often different from those used in the private sector. They can pose particular problems for road managers in allocating expenditures to particular activities or to particular sections of road.

Government accounting systems tend to be resource-based: sub-headings used represent items such as staff costs, capital expenditure, consumables, cost of utilities, and the like. Such systems may be useful in controlling general expenditure year-on-year, and can be used to provide conventional balance sheets. However, they do not provide a basis for relating expenditure to activities and works undertaken. This is the main concern for road maintenance management because, without this, it is not possible to determine unit costs that reflect accurately resources consumed in undertaking individual activities. Neither is it possible to make meaningful projections of cash-flow. A different approach to accounting is necessary if reliable information is to be obtained. This approach is a form of *management accounting*.

In addition, costing methods adopted by many governments do not take into account all of the items that are actually involved in an operation. These methods are designed to monitor expenditure against budget using a *cash accounting* system. In these systems, costs are written off in the year in which they are incurred, and no allowance is made for items procured that have a life beyond the present budget year. This makes it difficult to apportion costs of equipment and facilities to the specific activities in which they are used, resulting in the real costs of those activities being underestimated – sometimes by a considerable amount. For example, equipment costs are written down totally in the year of purchase, so any attempt by management to quantify and control the real cost of carrying out works is very difficult, since only the marginal or operating cost of the equipment is applied against the work. This is also the reason why it is normally difficult to compare government costs with those incurred by organisations operating in the private sector, such as contractors.

Furthermore, costings derived from cash accounting systems may exclude significant elements of cost, whereas the private sector must employ full cost accounting, including an element for return on capital employed (profit). Costs often not considered by government systems include equipment depreciation and down-time, supervisory staff expenses, office overheads and insurance, ownership costs of buildings, facilities, office equipment and furnishings, and the like. Using such an approach to costing would, clearly, underestimate the true cost of operations being considered in a financial analysis, and an *accrual* approach to accounting is necessary in this case to take into account items such as equipment depreciation and interest charges. There is a need for 'management' accounting rather than 'cost' accounting.

It is also particularly important for maintenance activities to consider the impact on the life of the facility and the resulting future cost streams. Life cycle analysis is needed to deal with this properly, and techniques for this are discussed later in this chapter. Again, traditional government accounting procedures are not well suited to dealing with this approach to costing.

4.3.3 Cost estimating techniques

Even with activity-based management accounting, it can still be difficult to provide realistic predictions of the total cash expenditure and time that will be necessary to complete the works (UMIST Project Management Group, 1989). This is because there are many variables that affect the cost of activities in different situations and circumstances. Traditionally, unit pricing techniques have been used for costing, but these have been shown to be deficient in several important areas. The use of analytical techniques and rigorous procedures of risk management are now recommended to produce realistic estimates of cost, particularly for major works. In particular, expected values of costs and timescales should reflect past experience of actual values achieved on completed works with similar characteristics.

Three basic estimating techniques are available to road managers for costing purposes:

- *Global*
  Overall method of costing, such as cost per kilometre of road works.
- *Unit rate*
  The traditional method of costing engineering works based on bills of quantities.
- *Operational*
  Resource-based costing, such as used by contractors for tender pricing.

These are listed in Table 4.1, together with the data required for their application. Each is described in more detail in the following sub-sections. A fourth technique, known as *worker hours*, is also sometimes considered. This technique is most appropriate for labour-based works, and is also discussed.

The global and unit rate techniques rely on historical data of various kinds, and comments on this aspect of each technique are given in the following sub-sections. However, certain dangers are associated with the use of historical data. These dangers are so critical that the following general warnings are worth making about the use of historical data in cost estimating.

Sample size

The data must be from a sufficiently large sample of similar work in a similar location and constructed in similar circumstances.

Base date

Cost data need to be related to a specific historical date, chosen with care. In the case of engineering work carried out over a period of time, the start date has traditionally been chosen for this.

Price indices

Having selected the relevant price base date, there remains the problem of updating the price to the base date for the estimate. The only practical method is to use an inflation index, but there may not be a sufficiently specific index for the work in question. If there is not, recourse to general indices is usually made. In any event, there is a limited length of time, which probably does not exceed five years, over which such updating has any credibility, particularly in times of high inflation.

Market effects

Overlying the general effect of inflation, is the influence of the 'market place'. This will vary with the type of works being undertaken. The state of the world economy at the price base date will require careful consideration before historical data can be credibly applied to a later, or future, date.

Table 4.1 *Cost estimating costs techniques*

Technique	Project data required	Basic estimating data required
Global	Size/capacity Location Completion date	Achieved overall costs of similar projects (adequately defined) Inflation indices Market trends General inflation forecasts
Unit rate	Bill of quantities (at least main items) Location Completion date	Historical unit rates for similar work items Preliminaries Inflation indices Market trends General inflation forecasts
Operational	Materials quantities Method statement Programme Key dates Completion date	Labour rates and productivities Plant costs and productivities Materials costs Overhead costs Labour rate forecasts Materials costs forecasts Plant capital and operating cost forecasts

Adapted from: UMIST Project Management Group (1989). Crown copyright 1989. Reproduced by permission of the Controller of H. M. Stationery Office.

4.3.4 Global techniques

This term describes the 'broadest brush' category of technique to give an overall cost. It relies on libraries of achieved costs of similar works related to the overall size or capacity of the asset provided. This technique may also be known as 'rule-of-thumb' or 'ball-park' estimating. Examples are:

- Cost per metre or kilometre of road resealing or overlay
- Cost per square metre of bridge deck areas or per cubic metre of mass concrete.

The technique relies entirely on historical data and, therefore, must be used in conjunction with inflation indices and a judgement of the market place influence to allow for the envisaged location and timing of the project.

The use of this type of generalised historical data for estimating is beset with dangers, especially inflation, as outlined generally above. The following specific dangers apply.

Varying definitions of what costs are included

- Engineering fees and expenses by consultants/contractors/client, including design, construction supervision, procurement and commissioning.
- Final accounts of all contracts including settlements of claims and any other payments.
- Land acquisition costs.
- Transport costs of materials.
- Financing costs.
- Taxes, duties, etc.

Varying definitions of measurement of the unit of capacity

- Metre/kilometre of road as an overall average, including or excluding *pro rata* costs of bridges, or estimates for these.
- Square metre of bridge deck area, including or excluding the cost of abutments; cubic metre of mass concrete in bridges, with height measurement from top of ground or top of foundations.

Not comparing like with like

- Differing levels of quality such as different pavement thicknesses for different levels of traffic.
- Different terrain and ground conditions such as roads across flat plains compared with mountainous regions.
- Different logistics depending on site location.
- Item prices taken out of total contract prices may be distorted by front end loading e.g. prices for hard currency items.

Inflation

- Different cost base dates; it is essential to record a 'mean' base date for the achieved cost and use appropriate indices to adjust to the forecast date required.

A scrutiny of all these dangers, especially the effects of inflation, must be made before any reliance can be placed on a collection of data of this type. It follows that the most reliable data banks are those maintained for a specific organisation where there is confidence in the management of the data. The wider the source of the data, the greater is the risk of differences in definition.

However, so long as care is taken in the choice of data, the global technique is probably as reliable as an over-hasty estimate assembled from more detailed unit rates drawn from separate unrelated sources and applied to 'guesstimates' of quantities.

4.3.5 Unit rate techniques

This technique is based on the traditional bill of quantity approach to pricing engineering work. In its most detailed form, a bill will be available containing the quantities of work to be constructed, measured in accordance with an appropriate method of measurement. The estimator selects historical rates or prices for each item in the bill using either information from recent similar contracts or published information (e.g. price books for civil engineering), or 'built-up' rates from their own analysis of the operations, plant and materials required for the measured item. As the technique relies on historical data, it is subject to the general dangers, as noted above.

When a detailed bill is not available, quantities will be required for the main items of work and these will be priced using generalised rates that take account of the associated minor items. Taken to an extreme, the cruder unit rate estimates come into the area of global estimates as described above (e.g. unit rate per metre of road).

The technique is most appropriate for repetitive work where the allocation of costs to specific operations is reasonably well defined and operational risks are easily manageable. It is less appropriate for works where the method of operation is variable and where the uncertainties of ground conditions are significant. It is also likely to be less than successful for engineering works in locations where few similar schemes have been completed in the past. In these cases, success depends much more on the experience of the estimator and their access to a well-understood data bank of relevant generalised rates.

Unit rates quoted by contractors in their tenders are not necessarily related directly to the items of work that they are pricing. It is common practice for a tenderer to distribute the monies included in the tender across the items in the bill to meet objectives such as cash-flow and anticipated changes in volume of work. Consequently, it follows that tendered bill unit rates are not necessarily reliable guides to costs for a particular work activity.

4.3.6 Operational costing techniques

Principles

This fundamental estimating technique compiles the total cost of the work from consideration of the constituent operations or activities revealed by the method statement and programme, and from the accumulated demand for resources. The advantages of working in current costs are obtained because labour, plant and materials are costed at current rates.

The most difficult data to obtain are the productivities of labour and equipment in the geographical location of the works, and especially in the circumstances of the specific activity under consideration. Claimed outputs of plant and equipment are obtainable from suppliers, but it is widely accepted that these need to be reviewed in the light of actual experience. Labour productivities will vary from site to site depending on management, organisation, industrial relations, site conditions, etc., and also from country to country. Productivity information is a significant part of the 'know-how' of a contractor and will naturally be guarded jealously.

The 'operational' technique is particularly valuable where there are significant uncertainties and risks. Because the technique identifies the basic sources of costs, the sensitivities of the estimate to alternative assumptions and methods can be investigated and the reasons for variations in cost appreciated. It also provides a detailed current cost and time basis for the application of inflation forecasts and hence the compilation of a project cash-flow.

In particular, the operational technique for estimating provides the best opportunity for identifying risks of delay as it involves the preparation of a method of construction and a sequential programme including an appreciation of productivities. Sensitivity analyses can be carried out to determine the most vulnerable operations and appropriate allowances included to reduce the effect of risks.

For each work activity, costings need to be built up for:

- Plant and equipment
- Labour
- Materials.

These are used to produce a *performance standard* for that particular activity. The performance standard provides a target, in both physical and cost terms, against which achievement can ultimately be measured.

Plant and equipment

For each item of plant and equipment, daily or hourly rates are built up by considering the costs of owning and operating each item of equipment. Ownership costs essentially comprise the cost of capital used to purchase the equipment, the cost of depreciation, and that of insurance. Operating costs comprise the repair and maintenance costs, and the cost of fuel and lubricants. Inclusion of ownership costs in the rates for plant and equipment is consistent with the accrual principle of accounting, in which costs and incomes are written to account in the period when they occur.

An example of the derivation of equipment rental rates is given in Table 4.2. This is based on the following methodology (Lantran and Lebussy, 1991):

Total annual depreciation	= (Renewal value – resale value + total lifetime interest charges) / (Service life)
Total maintenance and operation	= Annual cost of (parts + lubricants + mechanics wages)
Annual cost	= Total annual depreciation + total maintenance and operation + operator costs + general expenses
Total annual cost	= Annual cost + mark-up
Cost per day	= Total annual cost / average days utilised

Labour

Labour charge rates can be built up from salary or wage costs, social charges and overheads. Normally, a schedule can be provided for labour, which lists each employee and their appropriate charge rates. This schedule can be used to prepare detailed costings of works. For the preparation of performance standards and unit rates used for planning and estimating purposes, typical performance values are normally used for each class of labour based on average values.

An example of a labour charge rate calculation from a consultancy practice is given in Table 4.3.

Materials

Materials costing is relatively straightforward where materials are to be purchased on the open market. A schedule of materials can be prepared, listing for each:

- Description
- Unit of measurement
- Unit rate.

Costs can be determined relatively easily from the product of the quantity and the unit rate.

Where materials are manufactured by the organisation itself, such as for crushed aggregate, asphalt and the like, a more detailed breakdown of costs is appropriate. The following would need to be included:

- Procurement cost of raw materials (inputs).
- Cost of transport of raw materials to processing plant, including equipment and labour costs determined as above; will depend on haulage distance.
- Cost of processing, including equipment and labour costs; will depend on productivity of plant when working, and down-time when not being used.
- Cost of transport of processed material to site, as above.
- Cost of any wastage.

Table 4.2 Example of derivation of equipment rental rates

Equipment	Steel wheel roller	Rubber tyred roller	Multi-use truck	Bitumen sprayer	Bitumen tanker	Snow plough	Salt/sand spreader	Mechanical broom
Renewal value	25 000	110 000	130 000	85 000	75 000	11 000	45 000	10 000
Resale value	2 500	11 000	13 000	8 500	7 500	1 100	4 500	1 000
Life (years)	8	7	5	7	5	7	7	6
Days utilised	120	120	120	120	120	60	60	120
Operator cost	15 000	15 000	15 000	15 000	15 000	15 000	15 000	15 000
General expenses[1]	1 450	1 450	1 450	1 450	1 450	1 450	1 450	1 450
Interest[2]	15 000	57 750	48 750	44 625	28 125	5 775	23 625	4 500
Total annual depreciation	4 688	22 393	33 150	17 304	19 125	2 239	9 161	2 250
Spare parts[3]	2 500	15 571	20 800	9 714	12 000	1 257	5 143	1 333
Lubricants[4]	469	2 357	3 900	1 821	2 250	236	964	250
Mechanics wages[5]	625	3 143	5 200	2 429	3 000	314	1 286	333
Total maintn and operation	3 594	18 071	29 900	13 964	17 250	1 807	7 393	1 917
Annual cost	24 731	56 914	79 500	47 718	52 825	20 496	33 004	20 617
Mark up[6]	2 473	5 691	7 950	4 772	5 283	2 050	3 300	2 062
Total annual cost	27 204	62 606	87 450	52 490	58 108	22 546	36 304	22 678
Cost per day	272	522	729	437	484	376	605	189

Notes:
(1) General expenses are office and workshop rental, management and commercial personnel.
(2) Interest is 15 per cent per annum.
(3) Spare parts are 80 per cent of the equipment renewal value.
(4) Lubricants are 15 per cent of the equipment renewal value.
(5) Mechanics wages are 20 per cent of the equipment renewal value.
(6) 10 per cent mark up for risk and profit.

Developed using the method in: Lantran and Lebussy (1991).

Table 4.3 *Example of a labour charge rate calculation*

Staff and social charges as a percentage of salary	Per cent	Overheads	Per cent
Pension	9.50	Directors and admin	43.50
National insurance	14.60	Rent	12.60
Sick leave	1.90	General services	18.30
Holidays	12.70	Legal and audit	1.10
Staff welfare	2.00	Financial and bank charges	10.70
Other staff benefits	1.00	Depreciation, main and operating of assets	8.20
Sub-total	41.70	Insurance	6.90
		Business development costs	
		Time 4.90	
		Expenses 3.20	8.10
		Public relations	1.20
		Tax	4.70
		Sub-total	115.30

Staff member	Salary	Health insurance	Car	Annual total	Monthly total
J. Payne	36 000	592	2770	39 362	3280
H. Billington	30 000	592	2770	33 362	2780
G. Turner	25 000	592	—	25 592	2133
R. Davies	20 000	592	—	20 592	1716

Staff member	1 Basic remun/ month	2 Social charges (41.70% of 1)	3 O'head (115.30% of 1)	4 Sub-total	5 Fee (15.00% of 4)	6 O'seas allowance (40.00% of 1)	7 Charge rate per month
J. Payne	3280	1368	3782	8430	1264	1312	11 007
H. Billington	2780	1159	3206	7145	1072	1112	9 329
G. Turner	2133	889	2459	5481	822	853	7 156
R. Davies	1716	716	1979	4410	662	686	5 758

- Overheads of site, including any stockpiling costs.
- Mark-up for risk and profit.

Performance standards

A performance standard specifies the resource requirements for each activity to be carried out, and builds up a consistent description of the activity based on:

- A preferred and specified method of working
- Resources of equipment, labour and materials to perform the activity in accordance with the preferred method.

The performance standard relates the resources to an expected average daily accomplishment and, thereby, allows the calculation of unit rates for estimating purposes. These unit rates can be used to build up budget requirements, for both capital and recurrent activities, once work requirements have been identified.

An example of a maintenance performance standard is shown in Box 4.1 and examples of ranges of work outputs that can be expected are shown in Table 4.4. Unit rates can be derived from a performance standard by dividing the total daily resource cost by the average daily production. It can be seen that a performance standard presented in this way can be used directly as a work instruction to a road gang. It also provides a suitable format for reporting achievement.

Box 4.1	*Example of a performance standard*
ACTIVITY NAME	CODE
Open graded premix carpet Clause 509	
DESCRIPTION AND PURPOSE	UNITS
Laying and compacting an open graded wearing course carpet of 20 mm thickness in a single course composed of suitable small sized aggregates pre-mixed with a bituminous binder on a previously prepared base; in accordance with specification Clause 509.1	m^2
WORK METHOD	
1) Place signs and traffic control measures 2) Prepare base and carry out pre-patching (not included in this performance standard) 3) Ensure substrate is free from loose material by brushing manually or with a mechanical broom 4) Apply tack coat, Clause 503 (not included in this performance standard) 5) Prepare premix carpet in hot mix plant 6) Transport to site of laying in clean, covered trucks 7) Spread premix carpet with paver/finisher 8) Compact with 8/10 tonne smooth wheel roller 9) Apply seal coat conforming with Clause 513 (not included in this performance standard)	

(continued)

Box. 4.1 (*contd.*)

LABOUR	NUMBER	RATE	AMOUNT
Foreman/supervisor	2	108.00	216.00
Operator	7	93.00	651.00
Helper	5	88.00	440.00
Mechanic	1	108.00	108.00
Guard	2	88.00	176.00
Labourer	10	27.00	270.00
EQUIPMENT	NUMBER	Sub-total	1 861.00
Hot mix plant 20/30t/h	1	12 664.00	12 664.00
Paver/finisher 75t/h	1	12 208.00	12 208.00
Road roller 8/10t	2	7 208.00	14 416.00
Tipper truck 5/10t	2	4 448.00	8 896.00
Signs and traffic control	1	136.00	136.00
Power broom, tractor-mounted	1	1 864.00	1 864.00

MATERIALS	UNIT	QUANTITY	Sub-total	50 184.00
Aggregate 13.2 mm gauge	m³	29.7	350.00	10 395.00
Aggregate 11.2 mm gauge	m³	14.9	360.00	5 346.00
Bitumen in bulk	tonne	2.4	3 250.00	7 829.25
Diesel	litre	25.0	17.00	425.00

AVERAGE DAILY ACCOMPLISHMENT	APPROVAL	Sub-total	23 995.25
1 500 m²	Approved: Name/title: Date:	TOTAL RATE	76 040.25 50.69

Source: High-Point Rendel.

Table 4.4 *Examples of work outputs from activities*

Activity	Resource requirements			Output unit	Range of outputs
	Personnel	Equipment	Materials		
Clearing side drains by hand	4–10	Shovels, cutlasses, picks	—	m/worker-day	30–60
Clearing side drains by machine	2–3	Grader, shovels	—	km/day	4–7
Re-excavating side drains	2–10	Picks, shovels	—	m/worker-day	8–15
Clearing culverts	2–4	Shovels, head-pans/wheel barrows	—	no/worker-week	2–4

Activity		Tools	Materials	Units	
Minor repairs to culverts	2–4	Masons' tools	Cement, aggregate, sand	no/worker-week	2–10
Major repairs to culverts		To be assessed for each job		worker-day	—
Making culvert rings (1 m diameter × 1 m long)	4–10	Moulds, mixer, shovels	Cement, stone, sand, reinforcement	no/day	5–10
Grading unpaved surfaces	2	Grader, camber board, spirit level	—	pass-km/day[1]	20–50
Dragging unpaved surfaces	1	Tractor and drag	—	pass-km/day[1]	20–50
Patching bituminous surfacings	5–7	Pedestrian roller or hand rammers, brushes, picks, shovels, watering cans	Premix or gravel, bitumen, bitumen emulsion, chippings or washed gravel	m^3/worker-day	0.5–0.8
Filling gravel surfaces	5–7	Pedestrian roller or hand rammers, brushes, picks, shovels, watering cans	Gravel	m^3/worker-day	0.6–1.2
Filling earth surfaces and slopes	4–5	Hand rammers, brushes, picks, shovels	Selected earth	m^3/worker-day	0.9–1.5
Grass cutting by hand	2–10	Cutlasses	—	m^3/worker-day	300–800
Grass cutting by machine	1–2	Tractor/mower	—	km/day	10–20
Repairing and replacing traffic signs	2–3	Masons' tools, painters' tools, shovels	Cement, stone, sand, paint, reflective paint	no/worker-day	4–8
Road markings	2–4	—	Road paint	m/worker-day	50–200 (hand painting)
Stockpiling gravel by hand	10–20	Picks, shovels	—	m^3/day	450–500
Stockpiling gravel by machine	4	Bulldozer, loader	Gravel	m^3/day	300–350
Regravelling gravel surfaces	12–20	1 grader, 8 tippers, 1 loader, 1–2 water tankers, 1 roller	Gravel	m^3/day	300–350
Surface dressing	15–20[2]	1 distributor, 1 roller, 3 tippers, 1 gritter, 1 loader	Bitumen, chippings	lane-km/day	2.5–4.0

Notes: (1) 'Pass-km' is the actual distance the grader travels while working.
(2) + 10 additional if no loader is available + 10 additional if no gritter.

Adapted from: TRRL Overseas Unit (1987). Crown copyright 1987.
Reproduced by permission of the Controller of H. M. Stationery Office.

4.3.7 Worker hours techniques

This is most suitable for labour-based works where there exist reliable records of productivity of different trades per worker hour. The total worker hours estimated for a given operation are then costed at the current labour rates and added to the costs of materials and equipment. The advantages of working in current, as opposed to historic, costs are obtained.

The technique is often used without a detailed programme on the assumption that the methods will not vary from project to project. Experience has shown, however, that where they do vary (e.g. owing to the capacity of equipment used in conjunction with the works), labour productivities and consequently the total cost can be affected significantly. It is recommended that a detailed programme is prepared when using this technique. The prediction of cash-flow requires such a programme. The technique then becomes a special case of the operational costing technique.

4.3.8 Suitability of individual techniques

The different techniques that have been described differ in terms both of the quantity of their input data requirements and the accuracy of their outputs. It was noted in the discussion of management functions in Chapter 1 that, as the management process moves from planning, through programming and preparation to operations, the level of detail and accuracy of the data increase. The use of global estimating techniques at the planning stage is consistent with this concept. Similarly, as the name suggests, operational estimating techniques are appropriate at the operational stage of management. Hence, in general terms, as the management process moves through each of the four basic functions, it is appropriate for the cost estimating techniques used to change from global, through unit rates to operational.

A similar concept can be used to consider appropriate estimating techniques for new projects. International agencies, such as the World Bank and European Commission, use the concept of a 'project cycle' for planning and preparing loans for aid projects (Commission of the European Communities, 1993). The use of the word 'cycle' in this context is, perhaps, a misnomer because the process is essentially linear rather than cyclic. Typical steps in the project cycle were listed in Chapter 1:

1. Identification
2. Feasibility
3. Design and commitment
4. Implementation, operation and evaluation.

At the identification stage, the absence of all but the simplest definition of the project means that only the *global* technique can be applied. However, estimating organisations that regularly use operational techniques state that even their crudest overall data are recorded in such a way that the effects of the more obvious uncertainties can be allowed for at this early stage. Clearly the availability of a reliable, well-managed, global cost data bank together with associated broad brush analyses is an essential requirement for any organisation involved in the early identification of projects for inclusion in a forward financial programme.

The essential activity in the feasibility stage of a project is the consideration of alternatives. The most important characteristic of the estimating technique employed is, therefore, reliable comparability between the alternative projects, which may be numerous. The technique must also be usable with only preliminary data for the projects, as the conceptual design will normally be in its very early stages. The most appropriate techniques would be:

- Global
- Unit rate, using generalised unit rates for the main items of work.

As soon as sufficient data detail is available at the design and commitment stage, the first preference should be to use the operational technique. An exception may be labour-based projects, where the worker hours technique would be appropriate. In cases where, for whatever reason, there is insufficient time, funds or data available for the operational method, resort may have to be made to the unit rate technique. Operational costing should always be used at the implementation, operation and evaluation stage.

The similarity between different cost estimating techniques for each of the management functions and for the project cycle is shown in Table 4.5. However, when considering estimating techniques, the following factors should be kept firmly in the foreground:

- For all 'one-off' jobs, there is no credible alternative to operational estimating.
- Accuracy in all estimates depends heavily on a clear definition of scope, the extent of use of local information and on the definition of uncertainties and potential problems.
- There is considerable merit in using an alternative approach to prepare a 'validation' estimate; any differences between the main and validation estimates must be reconciled satisfactorily.
- An estimate submitted at any stage of a job should be subject to review.

Table 4.5 *Suitability of cost estimating techniques*

Management function	Project activities	Data detail	Accuracy	Cost of preparation	Estimating technique
Planning	Identification	Coarse/ summary	Relatively large tolerances	Relatively low	Global
Programming	Feasibility				Unit rate
Preparation	Design and commitment				
Operations	Implementation, operation and evaluation	Fine/detailed	Relatively tight tolerances	Relatively high	Operational

- It is recommended that all submitted estimates should include a carefully considered programme for the work; if this is omitted, there is a reduced likelihood that the effects of risk, delay and inflation will be considered.
- It is vital that any modifications to the estimate are backed up by a depth of study not less than the depth of the original estimator's own investigations.

4.3.9 Application of the approach to cost estimating

The approach to cost estimating described in the previous sub-sections is drawn from *A guide to cost estimating for overseas construction projects* (UMIST Project Management Group, 1989). This Overseas Development Administration guide deals more broadly with the management of risk in engineering projects, and has been applied successfully since 1989. The guide contains a pro-forma summary and explanatory notes which are highly recommended to those organisations that do not have their own cost estimating forms. Use of this approach has been successful in identifying the risks associated with project costs, and in identifying those risks that the client is in the best position to handle (currency fluctuations, for example), and those that are more appropriately transferred to the contractor. The main benefit from adopting this approach has been a substantial reduction in the number of projects over-running on time and budget.

4.4 Cost–benefit analysis

4.4.1 Purpose of the analysis

This section summarises the general principles for undertaking cost–benefit analysis. For more detailed discussion of the use of cost–benefit analysis techniques, particularly in the transport sector and roads sub-sector, reference should be made, for example, to Betz (1966), Van der Tak and Ray (1971), Bridger and Winpenny (1983), Adler (1987), Overseas Development Administration (1988) and TRRL Overseas Unit (1988a).

The purpose of carrying out cost–benefit analysis is primarily to ensure that an adequate return in terms of benefits results from committing expenditure. Any commitment of expenditure can be considered as making an investment, whether it be a capital project or an investment in maintenance. An additional purpose is to ensure that the investment option adopted gives the highest return in relation to the standards adopted, and the timing of the investment.

For economic appraisal, the assessment is made in terms of the net contribution that the investment will make to the country's economy as a whole. Thus, the analysis differs from that which would be undertaken by private companies in appraising commercial ventures in that it attempts to evaluate economic costs and benefits rather than financial ones. The essential approach is to use the *opportunity cost* of the investment as a measure of resource rather than market prices.

Each investment decision is unique and has features that prevent analysis following an identical pattern, although the same overall approach can usually be followed. It is normal to determine the costs and benefits that will be incurred over the analysis period if no investment is made, and compare these with the costs and benefits arising as a result of making an investment. Costs and benefits are compared as described in sub-section 4.4.3 to determine whether the investment is worthwhile, and to identify which is the best of the alternatives being considered.

The alternative in which no investment takes place is sometimes known as the *baseline* or *do nothing* case. However, it is unusual for future investment in such cases to be absolutely zero, as there is normally some kind of facility in existence which in the future will at least require some expenditure or minimum maintenance. In cases such as this, the *do minimum* alternative should be considered as the most realistic baseline case against which alternative investments should be compared. The choice of an appropriate *do minimum* case is an extremely difficult decision and has a very large influence on the size of economic return obtained. Considerable attention should therefore be given to its selection.

4.4.2 Prices

Resource costs

In order to carry out an economic analysis, it is necessary to make adjustments to costs and prices to ensure that they are all measured in the same units and that they represent real resource costs to the country as a whole.

Inflation

A first step in this is usually to remove the effect of inflation to enable values to be compared on the same basis over time. Costs and prices are normally expressed in constant monetary terms, usually for the first, or base year, of analysis. In most cases, it can be assumed that future inflation will affect both costs and benefits equally, so its effect can be ignored. However, there may be exceptions to this and, in these cases, different costs and prices will need to be assumed for different elements at different times in the analysis period.

Discounting

It is also necessary to factor costs and benefits to take account of the different economic values of investments made at different times during the analysis period. When money is invested commercially, compound interest is normally paid on the capital sum. The interest rate comprises inflation, risk and the real cost of postponing consumption. Thus, money used to invest in this situation could be invested elsewhere and earn a dividend. By using capital to invest in a particular item of work or in a project, the dividend is foregone and this should be taken into account in the analysis. To do this, all future costs and benefits are discounted to convert them to present values of cost (PVC) using the formula:

$$PVC = c_i / (1 + (r/100))^i$$

where c_i = costs or benefits incurred in year i
  r = discount rate expressed as a percentage
  i = year of analysis where, for the base year, $i = 0$.

Since the inflation element is dealt with separately (see above), and risk also needs separate treatment, the discount rate used may differ from market interest rates.

The value of the discount rate used will clearly have a considerable influence on the balance between the effect of investment costs, which are typically spent early in the project life, and that of benefits obtained in the future. Discounted benefits may exceed costs at one discount rate, but not at another. The choice of discount rate is therefore crucial to the outcome of an appraisal in many cases.

Shadow prices

Prices paid for goods and services may not reflect the real value of national resources because of distortions. These may be caused by governments fixing an unrealistic exchange rate (normally resulting in a 'black market' price for goods), because of control of imports and exports through quotas and subsidies, or because of the existence of monopolies or cartels that can 'fix' the price of certain commodities above their real resource value. The addition of taxes also distorts the real value because the taxation component represents a transfer of spending power from those purchasing the product to the government.

An economic cost–benefit analysis is seeking to improve the net economic benefit, or to improve the rate of economic growth, by considering how best to allocate real resources. Thus, in such an analysis, the effect of all price distortions must be removed. This is known as *shadow pricing*. The real resource cost of a commodity is found from:

  Resource cost = Financial cost – Taxes + Subsidies

This convention adjusts financial costs to take account of the fact that taxes and subsidies are transfer payments and do not reflect the cost of resources consumed in transport.

Note that, in this respect, economic analysis using cost–benefit methods differs from the *financial cost–benefit analysis* that would be undertaken, for example, by the promoter of a toll road. The financial analysis would be undertaken using market prices.

4.4.3 Comparison of alternatives

Exclusivity of alternatives

As noted earlier, cost–benefit analysis is concerned with the comparison of alternatives: often 'doing something' compared with 'doing minimum'. Sometimes the choice to carry out one alternative precludes the possibility of undertaking the other. This is known as *mutual exclusivity*, and most cost–benefit analysis is concerned with this situation. In effect, this is concerned with choosing the best way of making an investment. Sometimes, alternatives are not mutually exclusive, and the process then becomes one of ranking or prioritisation. This is discussed in Chapter 7.

Net present value

In order to determine whether an adequate return in terms of benefits results from making an investment, cost–benefit analysis must be carried out. This is normally done using the *net present value* (NPV) decision rule. This rule may also be used for helping to determine which investment option gives the highest return of those considered.

NPV is simply the difference between the discounted benefits and costs over the analysis period.

$$\text{NPV} = \sum_{i=0}^{n-1} \frac{b_i - c_i}{(1+(r/100))^i}$$

where n = the analysis period in years
i = current year, with $i = 0$ in the base year
b_i = the sum of all benefits in year i
c_i = the sum of all costs in year i
r = the planning discount rate expressed as a percentage.

NPV is a measure of the economic worth of an investment. A positive NPV indicates that the investment is justified economically at the given discount rate and, the higher the NPV, the greater will be the benefits or the lower will be the costs. Thus, the choice between investments should be based on NPV.

The NPV can only be calculated from a pre-determined discount rate which needs to be the same for each investment being compared. The NPV should only be quoted in conjunction with the discount rate that has been used.

A range of NPVs should always be quoted to reflect the range of scenarios being investigated by the cost–benefit analysis. It is also important to consider the results of the financial, social and environmental appraisals when deciding which is the best investment.

Internal rate of return

Although NPV is the correct decision rule for determining whether or not investments are worthwhile, its use is sometime impracticable. For example, an organisation such as the World Bank each year needs to appraise many hundreds of projects spread throughout a number of sectors in different countries. Comparing the NPV across all potential projects to see which would offer the best investment choice is not practical. Instead, organisations such as the World Bank use the *internal rate of return* (IRR) decision rule.

The IRR is the discount rate at which the present values of costs and benefits are equal; in other words, the NPV=0. Calculation of IRR is not as straightforward as for NPV, and is determined by solving the following equation for r:

$$\sum_{i=0}^{n-1} \frac{b_i - c_i}{(1+(r/100))^i} = 0$$

with notation as above.

The IRR gives no indication of the size of the costs or benefits of an investment, but acts as a guide to its profitability. The higher the IRR, the better the investment. If it is larger than the discount rate, then the investment is economically justified.

Organisations such as the World Bank set 'cut-off' rates of return for different countries. These are based on experience of what IRR values are necessary to justify Bank funding within country allocations that are likely to be available. As such, funding an investment whose IRR is above the cut-off rate provides a convenient mechanism for selection. Normally the cut-off IRR percentage rate is not less than the discount rate for the particular country, thus ensuring that the NPV is positive for investments that are selected.

NPV/cost ratio

One problem with the use of NPV is that, other things being equal, a large investment will normally have a larger NPV than a smaller one and, on the basis of this criterion, would always be chosen. This can cause difficulties when several potential investments are being compared under budget constraints. The problem can be overcome by considering the *NPV/cost* ratio. This ratio represents the magnitude of the return to be expected per unit of investment and is, therefore, a measure of the efficiency of an investment, since its value is increased either by:

- Increasing the size of the NPV, or
- Reducing the size of the cost.

Given a constrained budget situation, the most efficient investment is that with the largest NPV/cost. This investment should be undertaken first, followed in turn by those with successively lower NPV/cost ratios, until the budget is exhausted. This approach maximises the NPV that can be obtained from a limited budget. It enables several smaller investments, which in aggregate have a higher NPV, to be chosen over a single larger investment.

Choice of method

In most cases, the NPV and IRR will give consistent results and will identify the same investments as being worthwhile. If the use of IRR gives a different recommendation to NPV, then this should cast doubt on the analysis that has been undertaken. In general, where a government is using a target, or minimum return on capital, maximising NPV should be the criterion adopted. IRR is particularly useful when discount rates are uncertain. Normally both methods should be evaluated for an investment.

Whereas NPV and IRR are appropriate for appraising one-off investments, most of the need for cost–benefit analysis in road management concerns selecting those investments to make from a number that are possible, under conditions of budget constraint. In these situations, the NPV/cost decision rule is that which is most appropriate.

4.5 Life cycle costing

4.5.1 Investment choice

It should be emphasised that money spent on maintenance should be treated as an investment in the same way as for that spent on new construction. Hence, cost–benefit analysis is an appropriate tool for making decisions about maintenance expenditure. Also, as noted earlier, it is particularly important for maintenance activities to consider the impact on the life of the works and the resulting future cost streams. Thus, the application of cost–benefit principles to decisions about maintenance investment implies consideration of the concepts of life cycle costing.

Making choices on a rational basis requires the comparison of different levels of investment at the present time compared with their respective consequential future costs. The adoption of higher engineering or maintenance standards normally leads to higher investment costs, but may result in lower costs to the road administration in terms of future costs of maintenance and renewal. Such concepts of life cycle costs can be extended by considering on-going costs to road users. These consist of vehicle operating costs, time and delay costs, and the costs of road accidents.

If life cycle costs are not taken into account, investment decisions become subjective and dependent on the application of standards and intervention levels that are often themselves based on historical precedent rather than objective analysis. Life cycle considerations should also result in 'maintenance-friendly' measures being taken. The rational formulation of standards and intervention levels should also depend on life cycle cost considerations (Robinson, 1993), as should the selection of priorities for investment under budget constraint (see Chapter 7).

4.5.2 Road administration costs

Costs incurred by road administrations include the on-going disbursements for maintenance in all its forms to:

- Pavements
- Footways and footpaths
- Cycle tracks
- Drainage features
- Structures
- Signs, markings, signalling, road furniture, etc.

Some of these are fixed costs; others are variable, and depend on factors such as:

- The standard of the road concerned, from motorway to housing estate access.
- The geographic location within the country.
- The geotechnical environment through which the road passes, including topography, soils, etc.
- The degree of urbanisation surrounding the road, which will affect, in particular, structures.

- The sensitivity of the physical and socio-political environment through which the road passes, since this will influence the extent of measures necessary to mitigate any environmental damage.

4.5.3 Road user costs

These normally include:

- Vehicle operating costs, including both running and standing costs, during normal operation and during stop–start conditions when road works are being carried out.
- Time costs, including those for delays due to congestion and road works.
- Road accident costs, including those during the works.

These costs are interdependent. Vehicle operating costs depend on the number and type of vehicles using the road, the type of journey that they make, the geometry of the alignment and the condition of the road surface. Time costs are also related to road geometry, since a road built to high standards will allow higher travel speeds. Similarly, restricted-access roads reduce delays at junctions and in urban areas. Road accident rates are also related to the number of junctions and the skid resistance of the road surface. Surface characteristics also affect vehicle speed and operating costs, and these will also be a function of road maintenance policy and cost.

4.5.4 Life cycle cost models

The analysis of life cycle costs for roads assumes increased complexity because costs and the relationship between costs change over time. As time passes, roads deteriorate. As levels of deterioration increase, the need for road maintenance, and hence its cost, increases. However, as roads become rougher, vehicle operating costs will increase, as noted in Chapter 1. Travel time and accident rates may also change. Thus, road user costs will be affected by the condition of the road surface and this, in turn, will change over time as the road is maintained. The condition of the road surface is also affected by the design and construction standards, the traffic loading, the maintenance standards and the environment. The more vehicles that use the road, and the heavier their axle loads, the more quickly the road will wear out. The road will also wear out more quickly if it is subject to extremes of climate that may weaken the structure or cause erosion.

Thus, the road standards, the environment, the vehicles and the level of maintenance all have an effect on the cost and the changes in cost experienced by the road users. This can be considered schematically as shown in Figure 4.4. These basic considerations underpin the approach to modelling life cycle costs for roads.

It is possible to trace the use of life cycle costing of roads over one hundred years. Croney (1977) reports that, prior to 1870, engineers in the City of London used records extending back over 40 years to determine life cycle costs of stone sett pavements with the more commonly used water-bound macadam construction. More recently, the most well-known development of methods has been that of the HDM/RTIM series of models that is summarised in Table 4.6, although many other models

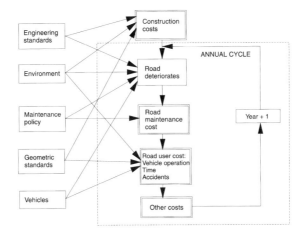

Figure 4.4 *Annual cycle of cost and road deterioration (Source:* Robinson, 1993)

Table 4.6 *Life cycle models for roads*

Date	Name	Developer	Comments	Reference
1968	HCM	MIT/World Bank	Based on literature survey of relevant work	Moavenzadeh *et al.* (1972)
1975	RTIM	TRRL/World Bank	New model extending HCM to incorporate results of field work in Kenya	Abaynayaka *et al.* (1977)
1979	HDM-II	World Bank/MIT	New model extending RTIM to increase analytical capacity	Harral *et al.* (1979)
1982	RTIM2	TRRL	Simplified version of RTIM also incorporating results of field work in the Caribbean	Parsley and Robinson (1982)
1985	micro-RTIM2	TRRL/University of Birmingham	Micro-computer version of RTIM2	Kerali *et al.* (1985)
1987	HDM-III	World Bank	New model extending HDM-II and incorporating results of field work in Brazil and India	Watanatada *et al.* (1987)
1989	HDM-PC	World Bank	Micro-computer version of HDM-III	Archondo-Callao and Purohit (1989)
1993	RTIM3	TRRL	Spread sheet version of RTIM2	Cundill and Withnall (1995)
1994	HDM Manager	World Bank	Menu-driven version of HDM-PC, later adding vehicle congestion realationships	Archondo-Callao (1994)
1998	HDM-4	International consortium led by the University of Birmingham	New windows-based model extending and updating all earlier versions	

120 Road Maintenance Management

have also been developed around the world (Robinson, 1993). Most operate using variations of the methodology illustrated in Figure 4.5, which reflects the changing costs over time, discussed above.

A life cycle approach to decision making is considered essential if the quality of decisions is not to be biased by short-term considerations leading to detrimental longer-term consequences and avoidable higher costs.

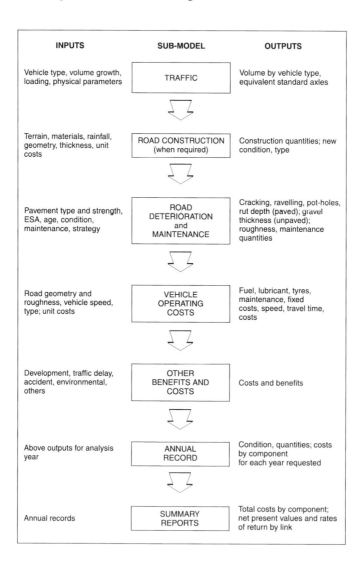

Figure 4.5 *Typical structure of life cycle cost model* (*Source*: Watanatada *et al.*, 1987)

5 Management Information

- Information and data
  - the need for information
  - information groups
- Data design
  - relevance, appropriateness, reliability and affordability
  - information quality
- Strategies for data collection
  - options available
  - cost–benefit analysis
- Inventory data
  - network referencing
  - item inventory
- Traffic data
  - traffic characteristics and volume
  - axle loading
- Pavement data
  - defects
  - manual and mechanised methods of pavement assessment
- Data management
  - computer-based data management
  - system selection and procurement

5.1 Information and data

5.1.1 The need for information

As noted in Chapter 1, management information sits at the heart of the management cycle. All management decisions rely on there being appropriate and up-to-date information to support them. Information is needed on items such as network details, traffic and axle loads, costs, road conditions, etc., which in turn require the existence of relevant data on which the information can be based. The need to assess physical condition, safety, level of service and efficiency of operation of road systems is widely recognised. In addition to knowing the characteristics of the existing system, it is becoming increasingly important to be able to predict the effects that proposed policies are likely to have in the future. Such predictive capabilities enable the decision-maker to test alternative courses of action to determine which policies and strategies will be the most effective in accomplishing the desired goals with the resources available.

Data are, therefore, needed to provide the basis for management decisions on such aspects as:

- Determining optimum road condition, and the maintenance strategies and expenditures needed to achieve this

- Determining optimum road condition within actual budget constraints
- Assessing current levels of road and bridge condition
- Determining appropriate levels of investment
- Prioritising capital improvements and investments in maintenance
- Simulating the effects of any improvements on the future condition and performance of the road system
- Estimating the cost of improvements
- Controlling on-going expenditures.

Management information of this nature provides the quantitative basis of dialogue between technical departments and ministries of finance during the fund allocation process. It can also provide the basis of a dialogue with elected representatives, road users, farmers and others, and for monitoring departmental performance and the meeting of policy objectives. Management information also needs to be understandable by all levels of staff within the road administration itself. The wide range of users of this information imposes a requirement that it should be presented in forms that are appropriate for the various audience requirements.

5.1.2 Information groups

It is convenient to discuss management information using the concept of *information groups*, such as shown in Table 5.1 This chapter discusses the data requirements for the first three of these:

- Road inventory
- Traffic
- Pavement.

Discussion of structures is beyond the scope of this book. Chapter 4 dealt with the data aspects of the remaining groups of:

- Finance
- Activities
- Resources.

But first, issues of data design and information quality are considered.

5.2 Data design

5.2.1 Data acquisition costs

The cost of data acquisition can be very high, and will often be the most expensive aspect of implementing and operating a road management system. System operation, itself, is likely to cost between 2 and 4 per cent of the maintenance budget provision. Many road administrations have fallen into the trap of collecting large amounts of data, but making only limited use of this for analysis and decision-making. Clearly, data should only be collected where their need can be justified. As such, it is essential that appropriate data design is undertaken if a cost-effective result is to be obtained.

Table 5.1 *Information groups*

Element	Aspects
Road inventory	Network/location Geometry Furniture/appurtenances Environs
Traffic	Volume Loadings Accidents
Pavement	Pavement structure Pavement condition
Structures	Structures inventory Structure condition
Finance	Costs Budget Revenue
Activities	Projects Interventions/treatments Commitments
Resources	Personnel Materials Equipment

Source: Paterson and Scullion (1990).

Detailed guidelines on data issues have been prepared by the World Bank (Paterson and Scullion, 1990). Their principles for data design are summarised here to provide a framework within which information for the roads sub-sector can be identified, stored and retrieved in an organised and efficient manner.

The following are the criteria that should be considered when selecting data items:

- Relevance
- Appropriateness
- Reliability
- Affordability.

These are discussed in turn.

5.2.2 *Relevance*

Every data item collected and stored must have a direct influence on the product or the output required from the system, which should already have been determined. Other data items which may be considered as desirable, interesting, or possibly useful in the future, should be omitted in favour of those that are essential, relevant and of immediate use.

Individual data items may be used for more than one function. For example, traffic data may be needed to enable a safety objective to be met, and may also be needed to enable pavement strengthening treatments to be designed. However, the use of common data items may differ in terms of level of detail, precision and scale between different applications. Thus, data relevance will depend on ultimate use, including the managerial level at which this use occurs. Relevance also implies that the data's dimensions and units comply with designated standards, particularly where national or international standards exist.

5.2.3 Appropriateness

Clearly, as the management system must be appropriate to the current needs and resources of the roads administration, then so should the data items. The technology and resources involved in acquiring, processing and managing the data should be appropriate to the administration's capacity for maintaining the equipment, conducting the surveys and sustaining the data processing.

The volume of data and the frequency of updating them are major determinants of the cost of operating the management system. The natural enthusiasm to collect and store every piece of data must be balanced pragmatically against the cost of collection, the time and cost for data storage and processing, and the consequential time required by managers to utilise the ensuing reports. Many systems have fallen into disuse because the resources required for data collection and processing have been too onerous.

Some types of data are collected at different times in a staged process, and the intensity and detail of measurement may differ between these stages, usually adding progressively more detail to the basic information acquired originally. Additional measuring techniques may need to be introduced as the level of detail required increases. For example:

- At the strategic *planning* stage
  – data on road condition would be collected across the whole network, but at a low sampling rate.
- At the project *preparation* stage
  – considerably more detailed information would be collected over the limited part of the network that comprised the project to refine the design and contract quantities.

The frequency with which data are updated is related closely to the issue of data volumes. Data that are out-of-date are usually irrelevant, particularly if the value changes rapidly. Trade-offs need to be made between precise data collected infrequently on the one hand, and less precise data, which are updated more frequently, on the other hand.

5.2.4 Reliability

The levels of accuracy and reliability required of data can vary considerably between different applications. For example, the accuracy of measurement for axle load enforcement must be high if prosecutions are to be obtained; whereas a lower accuracy of individual measurements is accepted for the purpose of pavement structural analysis, particularly if used in conjunction with a large sample of vehicles.

It is important to ensure consistency of data over time and between locations. A balance between the reliability of data and the certainty of the resulting product or outcome should be sought. For example, highly accurate, intensive sampling of entire networks, such as can be obtained using mechanised methods, may represent over-investment if the analysis method used is very general or approximate, or if the results are only to be used for broad *planning*. Reliability of data is determined from the following:

- *Accuracy*, which is the relationship between the measured and the 'real' value; this is defined by a combination of precision (the error associated with repeated measurements made at separate times or places, or by separate operators or instruments) and bias (the degree to which the mean measurement reflects the range and variability of all data points).
- *Spatial coverage*; for network-level *planning*, low intensity sampling is adequate whereas, for engineering design of projects at the *preparation* stage, intensive sampling is needed with full coverage of the project area.
- *Completeness* of data is important because missing items degrade the reliability of the product.
- *Currency* ensures that data are renewed with sufficient frequency to permit their effective use, recognising the need for data that change rapidly from year-to-year, or that have a large impact on the ultimate decision, to be kept up-to-date more than data that do not change so rapidly.

Reliability can also be affected by the data collection method used (Livneh, 1994). The aim should be for high levels of:

- *Repeatability*
  The ability of the *same* survey team or piece of equipment to record the same measured value on different occasions.
- *Reproducibility*
  The ability of *different* teams or equipment to record the same measured value as each other.

5.2.5 Affordability

The size and quality of all of the data items, and the associated data acquisition, must be affordable in terms of the financial and staff resources available to collect data and keep them current. The scope and quality of data are choices that must be weighed against the resources required to sustain them in the long term, and against the value of the management decisions that rely upon them.

Available resources and skills vary between road administrations. Generally, for small administrations, or wherever skills and resources are relatively scarce, simple and basic types of data, quality and collection methods must suffice. Where skills and resources are more abundant, a wider range of data, including the use of automatic collection methods, may be appropriate. Conversely, automatic methods may be appropriate in those organisations where resources are limited, particularly if the data are collected under contracts that are managed by a small number of skilled staff. In general, automatic data collection is preferred. Its use provides greater reliability, and results are more repeatable and reproducible. Survey prices are falling with the advent of cheap and relatively easy-to-use automatic data collection methods.

As administrations develop over time, and as their resources expand and improve, improved data collection methods can be adopted. But in order to do this, rational relationships between old and new forms of data are essential to preserve the value of historic data and enable its use, where required, for the analysis of trends.

5.2.6 Information quality

The concept of *information quality levels* (IQL), introduced by the World Bank, is particularly helpful when determining data requirements and designing data systems. The IQL concept recognises that different levels of data are needed for different levels of road management activity. It provides a framework for collecting and using data in a consistent manner when undertaking any particular activity. This helps to ensure that enough, and only enough, data are collected to enable appropriate decisions to be made, thereby saving unnecessary cost. The essential concepts of the IQL approach to data design are shown in Box 5.1, and the way in which its application was envisaged by the World Bank is shown in Table 5.2.

Considering again the four management functions discussed in Chapter 1, these may be grouped into *network* and *project* level activities, as indicated in Table 5.3. Describing the management functions in this way is helpful since it delineates more clearly the amount of data detail required for each function. As the management process moves from *planning*, through *programming* and *preparation* to *operations*, the level of data detail required can be seen to increase progressively, but to reduce in the extent of its network coverage. This feature can be used to assist the data design process by combining the functional levels of road management with the information quality levels shown in Table 5.2, to provide a rigorous basis for classifying data needs, as shown in Table 5.4.

Table 5.2 *World Bank proposals for the application of information quality levels*

Information quality level	Short description	Applications	Data collection
IQL-I	Most detailed and comprehensive	(1) Research (2) Operations (3) Advance design (4) Diagnosis	Short to limited lengths or isolated samples using specialised equipment; slow except for advanced automation

IQL-II	Detailed	(1) Preparation (design)	Limited lengths using semi-automated methods; or full coverage using advanced automation at high speed
		(2) Advanced programming	
		(3) Advanced planning	
IQL-III	Summary details with categorisation of values	(1) Programming	Full sample using high-speed, low accuracy semi-automated methods; or sample at slow speed; or processed from other data
		(2) Planning	
		(3) Basic design	
IQL-IV	Most summary	(1) Sector/network statistics	Manual or semi-automated methods, processed or estimated
		(2) Low-volume road design	
		(3) Simple planning and programming	

Source: Paterson (1991).

Box 5.1   *The concept of information quality levels*

IQL-I   Most comprehensive level of detail, such as would be used as a reference benchmark for other measurement methods and in fundamental research; would also be used in detailed field investigations for an in-depth diagnosis of problems, for detailed *preparation* and monitoring of sensitive *operations*, and for high class project design (e.g. using mechanistic or theoretical methods); normally used at project level in special cases, and unlikely to be used for network monitoring; requires high level of staff skills and institutional resources to support and utilise collection methods.

IQL-II   A level of detail sufficient for comprehensive *preparation* of works, activities and for standard design methods; for *planning*, would be used only on sample coverage, sufficient to distinguish the performance and economic returns of different technical options with practical differences in dimensions or materials; standard acquisition methods for project-level data collection; would normally require automated acquisition methods for network surveys and use for network-level *programming*; requires reliable institutional support and resources.

IQL-III   Sufficient detail for *planning* models and standard *programming* models for full network coverage; for project design, would suit elementary methods, such as catalogue-type with meagre data needs, and low-volume road/bridge design methods; able to be collected in network surveys by semi-automated methods or combined automated and manual methods.

IQL-IV   The basic summary statistics of inventory, performance and utilisation, of interest to providers and users; suitable for the simplest *planning* and *programming* models but, for projects, is suitable only for standardised designs of very low-volume roads; the simplest, most basic collection methods, either entirely manual or partly semi-automated; provides direct but approximate measures, and suit small or resource-poor agencies; alternatively, the statistics may be computed from more detailed data.

Source: Paterson and Scullion (1990).

Table 5.3 *Network and project level activities*

Activity	Management function
Network level	*Planning* and *programming* require information on the road system as a network of links with jurisdictional, functional, traffic demand and physical characteristics, and on resources and costs; the differences are that *planning* is involved with strategic (long-term) decisions, whereas *programming* is involved with tactical (mid-term) decisions; network-level data are used for the processes of forecasting and budgeting of resources for road works
Project level	*Preparation* includes the preparation of project packages containing specific lengths of road, etc., the primary aim at this level is the selection and design of treatment, identification of the exact location and extent of its application, and the costing, allocation and scheduling of funds for this
	Operations include those management activities operating in real time or over very short timescales; management activities include: maintaining roads and structures in functional operation through routine and emergency maintenance operations, operating and maintaining traffic control devices, project monitoring, equipment management, etc.

Table 5.4 *Data requirements for different management functions*

Management function	Time horizon	Spatial coverage	Typical information quality level
Planning	Long-term (strategic)	Network-wide	IQL-IV
Programming	Medium-term (tactical)	Network to sub-network	IQL-III/IV
Preparation	Budget year	Section or project (or scheme)	IQL-II/III
Operations	Immediate/very short-term	Sub-sections	IQL-I/II

5.3 Strategies for data collection

5.3.1 Options available

A variety of strategies can be used for collecting data on a regular basis, and different approaches will be appropriate for road networks of different characteristics managed by different road administrations.

In one approach, high level condition data (typically IQL-IV) are collected across the whole network each year. These data are used for planning and programming purposes. The programming exercise then collects more detailed data (typically IQL-III) on those sections where works are likely to be undertaken. More detailed data (typically IQL-II) are then collected on some of the sections for which designs are produced, or for which works are undertaken. As more detailed data are collected on any section, they replace or augment those collected in the earlier phase, with the result that data for different sections are held at different levels of detail at any time during the year.

In an alternative approach, relatively detailed data (typically IQL-II/III) are collected across part of the network on a rolling programme, perhaps with a cycle of three to five years. Each year, programming decisions are taken either using current data for individual sections, if available, or by projecting forward condition data from previous years to give an estimate of current condition across the whole network. Thus, all condition data tend to be stored at the same level of detail, but with a different age. In addition, more detailed data may be collected as part of the works design or execution process.

Other combinations of the above are also used, including the following examples:

- Annual data collection on the primary road network, whereas a cycle of data collection may be used on lower roads in the hierarchy
- Cyclic approach used for the whole network, but with data collected at a lower level of detail (IQL-III/IV)
- Cyclic collection methods can also be used without projection of condition.

Some road administrations collect detailed data annually across the whole network, although this approach is unlikely to stand up to investigations of cost-effectiveness.

All of the above strategies have different implications for the level of data detail recorded and stored.

5.3.2 Cost–benefit analysis

Ideally, a cost–benefit analysis should be undertaken to determine the survey strategy that is most appropriate, and the optimal level of data detail to be collected and used. For example, an analysis could be undertaken about the level of detail at which it was appropriate to collect information on road signs. Different levels of detail might be used to collect and store the following information:

Level 0: No information.
Level 1: The number and generic types of signs on a particular road section.

Level 2: Details of each individual sign on a section, in terms of its drawing number.
Level 3: Details of each individual sign on a section, with information stored on such attributes as its design, mounting position, construction, height, illumination, and location in terms of either
- Chainage and cross-sectional position or off-set, or
- Grid co-ordinates.

The results of such a cost–benefit analysis will differ for each road administration, and are likely to differ depending on such items as: the size and level of development of the network; geography, topography and climate; road hierarchy, traffic level, urban or rural location; and political or environmental sensitivity of the location on the network or the data item. The key issue is that the survey strategy adopted, and the amount and level of data that should be collected, should be subject to rigorous analysis to determine what is cost-effective for its subsequent application.

5.4 Inventory data

5.4.1 Types of road inventory

Clearly, it is not possible to manage any asset without first defining exactly what that asset is. There is, therefore, a need to define and reference the network to be managed. This is the purpose of a road *inventory*, and there are two main classes of data that are normally grouped together under this term:

- Network referencing – concerned with the location of the road and its appurtenances, together with its alignment and geometry.
- Item inventory – concerned with the physical attributes of the road.

Each is discussed in turn.

5.4.2 Network referencing

This involves breaking the network down into successively smaller links, segments and sections, each of which can be defined uniquely. It is appropriate to define these terms in the following way:

- *Route*
  A length of road between an origin and a destination.
- *Link*
  A length of road that is uniform in terms of the traffic volume that it carries.
- *Segment*
  A length of road that is uniform in terms of its geometric characteristics.
- *Section*
  A length of road that is uniform in terms of its physical characteristics; sections are normally the basic unit of a road network for road management purposes.

- *Sub-section*
  A subdivision of a section used for more detailed analysis of condition.

A set of routes, links, segments, sections or sub-sections can be grouped together to form a road *network*.

The process of referencing a network involves both logical and physical activities. For example, sections should be selected, on a logical basis, to have homogeneous characteristics, with the following typically being uniform for a section:

- Road class
- Traffic level (i.e. sections will not include any major junctions)
- Road geometry
- Pavement construction type
- Other administrative data deemed appropriate, such as administrative boundaries, speed limits, and the like.

Logical referencing normally requires that the section is identified by a unique and unambiguous reference or *label*. A label would typically consist of a road name, or number, and a length identified typically by a start and end chainage. It is possible to associate a string of text to a label in most systems to assist in identification in the field, such as 'from the junction with River Road to the speed limit sign on the outskirts of Half Way Tree'.

It is necessary to match the logical definition of the network with the physical referencing. Sections need to be identified physically on the ground to enable work crews to identify location. Many different systems of physical referencing are in use, ranging from concrete 'kilometre posts' to signs containing bar codes which can be read automatically by high-speed survey vehicles. Markings on the road surface are commonly used, again ranging from simple studs to machine-readable devices. However, the cost of marking in these ways can be relatively high, and many road administrations indicate the start and end of sections by chainage from a defined starting point.

The road network can, alternatively, be defined spatially using grid co-ordinates. Spatial referencing is particularly appropriate for certain route planning activities or for storing map-based information relevant to the network. The increasing availability, accuracy and cost-effectiveness of global positioning systems (GPS) will result in increased use of this mode of network referencing. Linear and spatial referencing systems can be combined by attributing grid co-ordinates to sections. Doing this enables map-based graphics to be produced, or for links to be made to *geographic information systems* (GIS).

In this context, it is appropriate to say something about *nodes*. These are used in connection with transportation planning, GIS, road accident recording and analysis, and for defining road geometry. The definition of a node is different in each of these cases. It is simplest to define a node as the point at which two sections meet. Other attributes of a node, such as implications about the existence of road junctions, or connectivity, should be defined additionally to this. The concept of nodes does cause considerable confusion because their meaning is interpreted differently by different people.

There are also two different approaches to defining road lengths for maintenance purposes: the first is to have relatively short fixed sections or sub-sections that are always inspected and treated as an entity; the second is to define sections to be as long as the physical consistency requirement will allow, but then to allow works to be carried out on any length within these sections depending on the precise location of defects: this is known as *dynamic sectioning*. The first approach has the advantage that it is easy to administer and that supporting management information systems required are simple and easy to design. The disadvantage is that treatments will not always be applied in the most cost-effective way because the length over which they are applied in a particular location will always be fixed. The second approach overcomes this problem, but at the expense of the system being more difficult to administer and the software required being more complicated.

Referencing a network is a surprisingly time-consuming activity if it is done systematically and in an unambiguous way. But good referencing pays dividends, and should be seen as providing the basis for all subsequent planning and management activities (Deighton and Blake, 1994).

5.4.3 *Item inventory*

Item inventories are those features of the road that remain reasonably constant over time. An inventory may consist of records of such items as the lengths and widths of carriageways and footways; and details of drainage, road signs and the like. These records are essential for allocating funds in an appropriate manner, and for calculating maintenance costs, putting work out to tender and for supervising work. Item inventory can be subdivided further into two groups:

- *Continuous item inventory*
  The model of a road that describes longitudinal features (carriageways, kerbs, shoulders, verges, footways, cycle tracks and medians).
- *Point item inventory*
  Other physical items, such as drains, road markings, street lights, signs, structures and the like, the position of which may be defined by its chainage.

The existence of a basic road inventory enables (Local Authority Associations, 1989):

- A more rational approach to the development and control of the maintenance budget.
- Improved pre-planning, contract formulation and control of work.
- Operational improvements to be identified, particularly for routine maintenance activities.
- A better understanding of the changing volume of demand, such as for the adoption of new roads or the provision of street lighting.
- The measurement of outputs, such as the cost of side drains cleaned, or the percentage of the network treated in any one year.

Inventory items relate to *features* of the road network. A *feature* is a basic element of the network, such as a carriageway, footway, bridge and the like. Individual inventory items are considered as *entities*, each of which has *attributes*. For example, in the United Kingdom Department of Transport's *Routine Maintenance Management System* (RMMS), the *entity* 'sign' has the *attributes* shown in Table 5.5.

Table 5.5 *Attributes and codes recorded for entity 'sign' in the UK RMMS system*

Attribute	Code recorded	
Item code	'SG'	
Cross-sectional position	1 = Left outside verge	9 = Right footway
	2 = Left footway	0 = Right outside verge
	3 = Left verge	Q = Acceleration splay
	4 = Lane 1	W = Left turning lane
	5 = Lane 2	E = Right turning lane
	6 = Lane 3	R = Bus lane
	7 = Lane 4	T = Crawler lane
	8 = Right verge	Y = Other
Chainage	(Numeric value)	
Identity code	(Alphanumeric value)	
Category	1 = Warning	5 = Matrix
	2 = Regulatory	6 = Message
	3 = Informatory	7 = Other
	4 = Hazard warning	
Illumination	1 = None	4 = Remote
	2 = Internal	5 = Reflectorised
	3 = External	
Diagram number	(Alphanumeric code)	
Mounting height	(Numeric value to nearest 0.5 metre)	
Mounting method	1 = Post	5 = Lamp post
	2 = Bridge	6 = Traffic signal
	3 = Gantry	7 = Other
	4 = Wall	
Size	• Width	
	• Height	
	(Numeric values to nearest 0.1 metre)	
Ownership	1 = Department of Transport	
	2 = County council	
	3 = District council	
	4 = Other	

Source: Department of Transport (1988). Crown copyright 1988. Reproduced by permission of the Controller of H. M. Stationery Office.

The list of the items collected and stored in RMMS is shown in Table 5.6. This represents a comprehensive list of inventory items that will be expensive and time-consuming to keep up-to-date. For many road management applications, road administrations will only collect and store sub-sets of the items in this list. A sub-set used for low volume roads is given by the American Society of Civil Engineers (1992). A balance needs to be struck between the cost of storing data and the value obtained from the use to which they will be put. The *information quality level* concept is of particular relevance in this respect, and its application to road inventories is shown in Table 5.7. It may be appropriate, when designing an inventory system, for a detailed analysis to be undertaken of the use to which recorded data will be put, and the value

Table 5.6 *UK RMMS inventory items*

Geometry	Drainage	Features	Furniture
Carriageway type	Balancing pond	Bridge over	Safety bollards
Carriageway surface type	Catch-pit	Bridge under	Communications cabinet
Carriageway length	Channel	Central island	Emergency telephones
Carriageway width	Culvert	Central reserve	Fences and barriers
Footway surface type	Ditch	Cross-over	Hedge
Footway length	Filter drain	Cycle track	Lighting point
Footway width	Grip	Embankments and cuttings	Man-holes
Gradient	Piped grip	Kerb, safety kerb	Reference marker point
Radius of bend	Gulley	Lay-by	Pedestrian guardrails
	Metal grid channel	Pedestrian crossing	Road markings:
	Drainage kerb ('beaney block')	Verge	• Hatched
		Tapered kerb	• Longitudinal
		Subways	• Transverse/ special
		Public forecourts	Road studs
			Safety fence
			Signs
			Traffic signals
			Tree
			Inductor (detection) loops
			Pavement lights
			Town/village signs
			Pedestrian signs
			Bus stops
			Neighbourhood watch signs
			Sign stones
			Street name plates
			Parking meters
			Level crossing

Source: Department of Transport (1988). Crown copyright 1988. Reproduced by permission of the Controller of H. M. Stationery Office.

to be obtained from collecting certain items at given levels of accuracy. Increasing the accuracy of data collection may increase costs dramatically.

Table 5.7 *The use of information quality levels for road inventory*

Data group	IQL-I	IQL-II	IQL-III	IQL-IV
Length Vertical alignment	(1) Centre line elevation @ 10–100 m centres Av. absolute gradient	(1) Max. gradient Av. absolute gradient Length non-compliant	(1) Rise + fall[1]	(1) Gradient[2]
Horizontal alignment	For each segment: • Transition points • Curve parameters • Sight distance	Maximum curvature Av. curvature Length of tangent Length non-compliant Number non-compliant Min. design speed	Curvature[1]	Curvature[2]
Transverse profile	Camber Superelevation Pavement width Shoulder width[3] Shoulder types[3] Lane widths[3] No. of lanes Lane types[3] Median types[2] Median width Kerb type[2] Kerb height (L, M, R)	Camber Superelevation Pavement width Shoulder width Shoulder type Lane width[2] No. of lanes — Median type Median width Kerb type Kerb height	— — Width[1] Shoulder width[1] Shoulder type[2] Lane width No. of lanes — Median[2] — Kerb[2] —	— — Width[2] Shoulder[2] — — No. of lanes — — — — —
Formation	Type[1] Height (L, R) Width Side slope (L, R)	Type[2] Height[2] Width[2] Slope[2]	Type[2] — Width —	Type[2] — — —

Notes:
(1) Field measurement
(2) Pre-defined range options for estimation
(3) Every point in data set
L, M, R Left, middle, right
Av. Average
Max. Maximum

Source: Paterson (1991).

5.5 Traffic data

5.5.1 Types of traffic information

Traffic data are a measure of the use of the road, and therefore provide important information about the needs of the road user or 'customer'. Traffic data are also necessary to enable projection of road performance. Various types of traffic information are needed to assist with road management:

- Traffic characteristics and volume
- Axle loading.

Accidents are also a form of traffic data (see Table 5.1) but, since the subject of road safety is large and complex, it is not appropriate for consideration within the scope of this book. Further information can be obtained from documents such as those published by IHT (1990) and TRRL Overseas Unit (1991).

5.5.2 Traffic characteristics and volume

Traffic characteristics

For the purposes of design and the evaluation of benefits, the volume of current traffic needs to be classified in terms of vehicle type. An example of a typical classification is shown in Table 5.8. In order to assess benefits, it is also necessary to separate traffic into the following three categories, since each is treated differently in a cost–benefit analysis (TRRL Overseas Unit, 1988a):

- *Normal traffic*
  Traffic which would pass along the existing road if no investment took place, including normal growth.
- *Diverted traffic*
  Traffic that changes from another route (or mode) to the road, but still travels between the same origin and destination (this is termed *reassigned* traffic in transport modelling).
- *Generated traffic*
  Additional traffic which occurs in response to the provision or improvement of a road (this includes *redistributed* traffic as defined in transport models).

Traffic flow

Estimates of traffic flow along road sections are needed for most aspects of planning and management. The level of traffic will, for example, influence the road standards of geometric and pavement design, and the maintenance standard in terms of frequency of maintenance activities.

The first step in assessing demand is to estimate baseline traffic flows or, in other words, to determine the traffic volume actually travelling on the road at present. The estimate normally used is the *annual average daily traffic* (AADT), classified by vehicle category. This is defined as the total annual traffic in both directions divided by 365. Estimates of AADT are normally obtained by recording actual traffic flows over a specific shorter period than a year, and results are scaled to give an estimate of AADT. Both manual and automatic methods of counting can be used for this, and each is appropriate in different situations. Both methods are prone to inaccuracy: automatic methods because of difficulties of setting vehicle sensors to record vehicles or axles correctly, and manual methods because of human error. An example of the large errors typical of manual counting is illustrated in Table 5.9.

Table 5.8 *Example of a traffic classification*

Class	Description
Cars	Passenger vehicles seating not more than five persons, station wagons and taxis
Light commercial vehicles	All two-axled vehicles with single rear tyres not included as *Cars* or *Mini-buses*
Mini-buses	Mini-buses with 9–15 seats
Buses	Purpose-built bus with more than 15 seats
Light trucks	Two axle truck, petrol driven, and with twin rear tyres
Medium trucks	Two axle diesel truck with twin rear tyres
Heavy trucks	Three axle diesel truck with twin rear tyres
Articulated vehicles	Multi-axle articulated tractor and trailer

Traffic counts carried out over a short period as a basis for estimating the AADT can produce estimates which are subject to large errors because traffic flows can have large hourly, daily, weekly, monthly and seasonal variations. The daily variability in traffic flow depends on the volume of traffic, increasing as traffic levels fall, and with high variability on roads carrying less than 1000 vehicles per day, as shown in Figure 5.1. Traffic flows vary more from day-to-day than week-to-week over the year, so that there are large errors associated with estimating annual traffic flows (and subsequently AADT) from traffic counts of a few days duration. For the same reason, there is a rapid fall in the likely size of error as the duration of counting period increases up to one week, but there is a marked decrease in the reduction of error for counts of longer duration, as shown in Figure 5.2. Traffic flows also vary from month-to-month, so that a weekly traffic count repeated at intervals during the year provides a better base for estimating the annual volume of traffic than a continuous traffic count of the same total aggregate period. Traffic also varies considerably through the day, but this is unlikely to affect the estimate of AADT provided sufficient and appropriate hours are covered by the daily counts.

The key issue is that, recognising that estimates of traffic flow are subject to large variability, the amount of effort put in to reducing the variability should depend on the use to which the data will be put in terms of the ultimate management function. Some recommended traffic data detail for different IQL values is shown in Table 5.10.

Table 5.9 *Different values obtained by 10 observers during a one hour manual traffic count*

Vehicle class	Number of vehicles counted by each observer (A to J)										
	A	B	C	D	E	F	G	H	I	J	Range
Cars	176	180	299	198	361	300	396	446	293	338	176–446
Light commercial vehicles	268	245	102	268	52	125		9	149	104	9–268
Mini-buses	14	13	13	16	31	24	11	31	31	15	11–31
Buses	48	50	48	60	40	41	34	32	31	49	31–60
Light goods vehicles	13	13	16	14	22	13	24	11	17	12	11–24
Medium goods vehicles	0	0	10	0	3	6	15	13	11	12	0–15
Heavy goods vehicles	0	0	2	0	0	2	0	0	2	0	0–2
Articulated vehicles	0	0	0	0	0	0	0	0	0	0	0
Total	519	501	490	556	509	511	480	542	534	530	490–556

Note: Vehicle classification is that defined in Table 5.8.
Source: Survey undertaken by graduate students at the University of Dar es Salaam on 21 April 1995.

Management Information   139

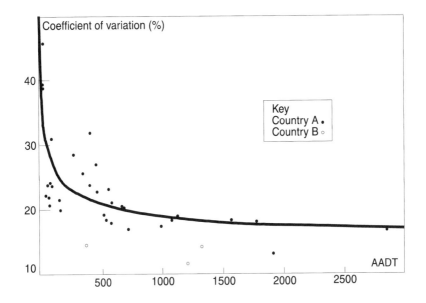

Figure 5.1 *Relationship between daily variability and traffic flow* (*Adapted from*: Howe 1972. Crown copyright 1972. Reproduced by permission of the Controller of H. M. Stationery Office)

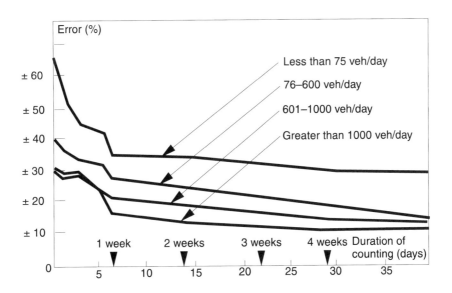

Figure 5.2 *Errors in AADI estimates from random counts of varying duration* (*Adapted from*: Howe, 1972. Crown copyright 1972. Reproduced by permission of the Controller of H. M. Stationery Office)

Table 5.10 *Traffic volume and axle loading data for different IQL values*

Data group	Information quality level			
	I	II	III	IV
Volume				
Total volume	AADT, seasonal ADT, hourly and short-term flows	AADT and seasonal AADT	AADT and seasonal factor	AADT range
Directional characteristics	By direction and lane	By direction, average heavy vehicles per lane	None	None
Composition	By vehicle class	By vehicle class	By 2–3 categories (e.g. heavy, bus, light)	Proportion of heavy vehicles
Loading				
Axle loading	Axle load spectrum	Average ESA per vehicle class, maximum axle load	Link/region average ESA per vehicle class	Regional average ESA per heavy vehicle
Gross vehicle mass	Spectrum by vehicle class	Average and maximum by class	None	None
Tyre pressure	Average and maximum by vehicle class	None	None	None

Source: Paterson (1991).

Traffic growth

Estimates must also be made about how this traffic will grow in the future, as many road management decisions are very sensitive to traffic forecasts. Forecasting growth reliably is notoriously difficult, especially when considering the variability likely to be present in estimates of current flows, as discussed above. Even in industrialised countries with stable economic conditions, large errors can occur but, in countries with developing or transitional economies, the problem becomes more intractable.

Different methods of forecasting tend to be used depending on the type of traffic being considered. These can be summarised by the following:

- Normal traffic
  - based on extrapolation of historical time-series data for traffic growth, fuel sales, GDP, or other relevant parameters, but taking into account any specific local circumstances.
- Diverted traffic
  - estimates normally based on origin and destination surveys.

- Generated traffic
  - normally based on 'demand relationships' which indicate the likely increase in traffic level for different levels of cost savings; generated traffic is particularly difficult to forecast.

It is beyond the scope of this book to go into the subject of traffic forecasting in detail.

5.5.3 Axle loading

The deterioration of pavements caused by traffic results from both the magnitude of the individual axle loads and the number of times that these loads are applied. Factors such as tyre pressure and wheel configuration are also important, but are difficult to deal with by the road maintenance management organisation. For pavement design and maintenance purposes, it is therefore necessary to consider not only the total number of vehicles that will use the road, but also the axle loads of these vehicles. To do this, the axle load distribution of a typical sample of vehicles using a road must be measured. The axle loads can be converted using standard factors to determine the damaging power of different types of vehicle. This damaging power is normally expressed as the number of *equivalent standard axles* (ESA), each of 80 kN, that would do the same damage to the pavement as the vehicle in question. This *damaging power* is termed the vehicle's *equivalence factor* (EF) and the design lives of pavements are expressed in terms of ESAs that they are designed to carry.

The relationship between the vehicle's equivalence factor and its axle loading is normally expressed in terms of the axle mass in kilogrammes (kg), rather than in terms of the force that they apply to the pavement in kilonewtons (kN). The relationship is normally considered to be of the form:

$$EF = \sum_{i=1}^{j} \left[\frac{axle_i}{8160}\right]^n$$

where $axle_i$ = mass of axle i (kg)
  j = the number of axles on the vehicle in question
  n = a power factor that varies depending on the pavement construction type, subgrade and assumptions about 'failure' criteria, but with a value typically of around 4.0

and the standard axle load is taken as 8160 kg.

This relationship is illustrated in Figure 5.3, where a power factor of 4.0 has been used.

The only effective way to determine the damaging effect of traffic on different roads is to measure the complete spectrum of axle loads, and to calculate the consequential damage in terms of standard axles. Axle load surveys can be carried out using a variety of types of equipment, ranging from small portable weighbridges, through 'weigh-in-motion' systems, to fixed static weighing platforms.

In the past, it has been customary to assume that the axle load distribution of heavy vehicles will remain unchanged for the design life of a new or strengthened pavement. However, more recently it has become clear that there is a tendency for steady growth over time of vehicle axle loads, and forecasts of this are needed to avoid underestimating future pavement damage. There are also examples where the introduction of a fleet of new and different vehicles can radically alter the axle load distribution on a particular route in a short time. Such events cannot normally be forecast and, hence, extrapolation from existing axle load surveys cannot provide for this sort of eventuality. Regular traffic surveys should highlight such situations, and new axle load surveys should then be carried out, as appropriate.

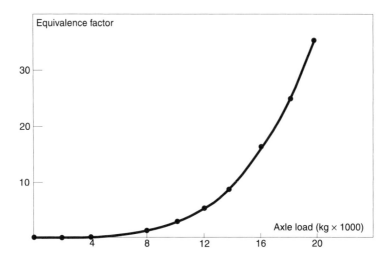

Figure 5.3 *Relationship between equivalence factor and axle loading*

As with traffic flows, the quality level of axle load data collection should vary depending on the road management function being undertaken. Table 5.10 gives levels of detail considered appropriate to different IQL values.

5.6 Pavement data

5.6.1 Defects

The network definition and inventory items provide basic information about the network to be managed. These data do not change, or change very slowly, over time. Information about the condition of the network does change, sometimes rapidly over time. The comparison of measured road condition with pre-defined standards, or intervention levels, provides a basic statement of shortfall in serviceability which can be translated into maintenance need. This process of determining need is discussed further in Chapter 6. Pavement condition is assessed by observing or measuring defects, and it is convenient to group these in terms of the following parameters (Peterson, 1987).

Roughness

This is sometimes identified as 'riding comfort', and is a measure of the longitudinal unevenness of the road. It is a major determinant of changes in vehicle operating cost and is, therefore, probably of greatest concern to the road user. Roughness is normally measured in units on the *international roughness index* (IRI) (Sayers *et al.*, 1986), and is normally assessed using mechanised means.

Surface distress

This is of most concern to maintenance personnel, and generally shows up as rutting, cracking, spalling and the like. Once visible signs of distress appear, then more expensive corrective maintenance is likely to be required to prevent more major forms (e.g. untreated cracks can lead to water ingress of the road base, resulting ultimately in a structural defect). Distress is normally assessed by visual methods or manual measurement although, increasingly, mechanised systems are becoming available for certain types of defect.

Structural capacity

As traffic accumulates, so does structural damage to the pavement layers. Some of this, such as rutting or cracking, is seen at the road surface; other damage, such as road base fatigue or sub-base soil contamination, is not. Indirect methods of assessment are usually employed, most of which are based on measurement of surface deflection under a dynamic load applied either by a moving vehicle or a dropped weight.

Pavement texture and friction

These give an indication of the safety of the road through their ability to prevent the skidding of vehicles. Measurement may be divided into two categories:

- Texture
- Wet-skidding resistance.

These are normally made with manual or vehicle-mounted instruments, which are used to identify potentially hazardous situations requiring corrective action.

Defects are assessed using condition surveys. These can be undertaken in a number of ways ranging from high-speed to slow-speed, mechanised to manual; destructive and non-destructive, and the like. Pavement assessment is discussed in detail by others, including Snaith (1985), OECD (1990), Haas *et al.* (1994) and Shahin (1994). As with inventory and traffic data, the IQL concept can be used to determine appropriate levels of data to be used in conjunction with different road management functions, and this is again discussed by Paterson and Scullion (1990). The range of pavement assessment methods is summarised in Table 5.11. In order to obtain a full view of defectiveness, several different measures may need to be collected. They are dis-

Table 5.11 Methods of measuring pavement condition

Defect	Examples of methods[1]	
	Manual	Mechanised
Roughness	Not really appropriate, although visual assessment can be used	*Profile measuring devices* • Inertial profilometers, including APL Profilometer, Automatic Road Analyser (ARAN), FHWA Profilometer, GM/Law Surface Dynamics Profilometer, Road Surface Tester (RST) • Laser profilometers, including High-speed Road Monitor (HRM), High-speed Survey Vehicle (HSV), Multi-function Road Monitor (MRM) *Surface-response devices* • Vehicle-mounted, including bump integrator, Law Roughness Surveyor, Mays Ride Meter, NAASRA Meter • Towed units, including BPR Roughometer, towed fifth-wheel bump integrator
Surface distress	Visual assessment using simple equipment	*Rutting* ARAN, RST, University of Birmingham Data Collection Van *Cracking* GERPHO and derivatives
Structural adequacy	*Destructive* Direct sampling and materials testing *Non-destructive* Benkelman beam *Semi-destructive* Dynamic cone penetrometer	*Impulse load deflection* Falling weight deflectometers: Dynatest, KUAB, Phoenix *Moving load deflection* Deflectograph, curviameter *Steady-state vibratory load deflection* Dynaflect, Road Rater
Pavement texture and friction	*Texture depth* Sand patch test *Skidding resistance* Skid pendulum	*Texture depth* Laser texture meter *Skidding resistance* • Sideways force coefficient using SCRIM • Braking force coefficient using the Grip Tester

Notes:
(1) Descriptions of these methods can be found in standard texts, such as Haas *et al.* (1994) and Shahin (1994).

cussed, in outline only, in the following two sub-sections under the headings of manual and mechanised methods.

5.6.2 Manual methods of pavement assessment

Roughness

The measurement of roughness is not practical by manual methods, although visual assessment of this can be used in the absence of any other means being available.

Surface distress

Certain types of defects are still difficult to measure by machine, and visual methods are the only ones that can be used. In particular, cracking is still difficult to measure reliably by mechanical means. The aim of visual inspections is to identify defects occurring on the road surface and to measure their severity and extent.

Structural capacity

Structural capacity measurements can make use of techniques that are:

- Destructive
- Non-destructive
- Semi-destructive.

Destructive methods require that the pavement structure is penetrated by digging a hole so that measurements can be taken of the thickness and properties of the materials in the various pavement layers and of the subgrade supporting the pavement. Typical measurements include density, particle size distribution and CBR.

The principal non-destructive method used is that of pavement *deflection*. This a measurement of the elastic response at the surface of a pavement subject to a known load. It is assumed that the magnitude of the elastic response may be used to determine the overall condition and weaknesses of the pavement as a whole. It is claimed that analysis of deflection results enables weaknesses in individual pavement layers to be identified, and allows prediction of the remaining life of the pavement to be made. Deflection can be measured manually using a *benkelman beam*. There are slight differences in the method of operation in different countries, and the weight of the wheel load differs. But the method is widely used.

A 'semi-destructive' method makes use of a *dynamic cone penetrometer*. This requires that a 20 mm hole is drilled through a bound pavement surface to enable the instrument to reach the lower unbound pavement layers. A standard cone is then driven through the pavement layers from the surface by repeated blows of a standard falling weight. The relative strength of each layer is quantifiable in terms of the penetration of this cone for each repetition of load. As the size of the hole is relatively small, its reinstatement is a simple matter.

Pavement texture and friction

For safety reasons, it is necessary to quantify the ability of the surface to prevent skidding. The first approach to assessing this is to measure the macro-texture depth of the road surface, typically with a *sand patch test*. The second is to measure the coefficient of friction mobilised in the presence of water between a road tyre and the surface of the pavement, known as the *skidding resistance*. The value of skidding resistance varies with speed, and any measurement quoted must be associated with a particular speed. A further complication is that considerable seasonal variation occurs because of the accretion of various compounds on to the road surface from passing traffic. During periods of high rainfall, these are flushed from the surface to allow a reasonably high degree of contact between the tyre and the road surface. But in prolonged periods of low rainfall, this layer can form a low friction interface with the tyre, resulting in low skidding resistance when it does rain. The skidding resistance may be measured at isolated points with a *pendulum tester*.

5.6.3 Mechanised methods of pavement assessment

Roughness

There are two main classes of method for measuring roughness (Shahin, 1994):

- Profile measurement
- Surface-response.

Devices for measuring the absolute profile of the road include:

- Inertial profilometers
- Laser profilometers.

Examples of different devices are listed in Table 5.11. Inertial profilometry requires the following four basic components (Haas *et al.*, 1994):

- Device to measure the distance between the vehicle and the road surface.
- An inertial referencing device to compensate for the vertical movement of the vehicle body.
- A distance odometer to locate the profile points along the pavement.
- An on-board processor for recording and analysing the data.

The various machines use different types of transducers for measuring the distance between the pavement surface and the vehicle. The inertial referencing device is usually either a mechanical or electronic accelerometer mounted to represent the vertical axis of the vehicle. As the vehicle passes along the pavement, vertical accelerations of the vehicle body are measured with the accelerometer. The accelerations are integrated twice to quantify the vertical movement of the vehicle body. These movements are added to the distance measurements to obtain elevations of the pavement profile. Some devices record the actual profile of the pavement surface, while others process the data on board and produce only a summary 'roughness' statistic.

As the name suggests, laser profilometers use lasers to measure distances to the pavement surface. As the vehicle moves along the road, laser-reflection methods are used to measure the distance between a sensor and the pavement surface at the position of the preceding sensor measurement. Thus, back-sights and fore-sights, as used in conventional surveying techniques, are recorded continuously and the profile of the road is determined.

The two types of profilometer can operate at high speed and are capable of determining accurate longitudinal profiles. However, they tend to be sophisticated and expensive, and this has led to the wide-spread use of surface-response instruments. These are simpler and cheaper, but do require calibration (Sayers *et al.*, 1986). Again, there are two classes of these:

- Vehicle-mounted
- Towed units.

Particular examples of these are given in Table 5.11.

Surface distress

The automatic measurement of surface distress has proved to be a difficult challenge for instrument designers. Successful work has been done on the measurement of transverse profile, including rutting, using ultra-sonics and laser technology. The measurement methods are similar to those for longitudinal profile, described earlier, but the sensors are mounted transversely, usually at the front of the survey vehicle. The two methods have similar resolution, but ultra-sonics have, in the past, had a price advantage. However, laser technology is now becoming cheaper.

Cracking has proved more problematic. The bulk of the successful work to date has been concerned with the automatic collection of data, rather than its analysis by automatic means. Various devices have been used to record images of the road surface under constant lighting conditions using a vehicle travelling both at creep and normal road speeds. Early versions of these used photographic images which were analysed manually to determine the severity and extent of surface defects. More recent developments have used 'shuttered' video cameras instead of photographic equipment. Research is now also advanced at a number of centres into image processing for road condition surveys which has the aim of automating the crack detection and analysis process.

Structural capacity

Mechanised methods of measuring structural capacity use non-destructive techniques: in addition to the measurement of rutting, described above, assessment of the structural capacity of pavements normally makes use of *deflection* methods. Examples of the three principle types are given in Table 5.11. These operate in different ways.

Impulse load deflection

Instruments in this class are normally known as *falling weight deflectometers* (FWD). They work from a stationary position on the pavement by dropping a weight onto a spring from a known height to impart a sinusoidal pulse to the road surface: this is intended to simulate the effect of a passing wheel load. Deflection measurements are made at the centre of the loaded area and at a number of positions off-set longitudinally. In this way, the profile, or 'deflection bowl' is obtained as well as the absolute maximum deflection. With this, it is claimed to be possible to 'back-analyse' pavement structures with computer software using an elastic or quasi-elastic technique to determine the structural state of constituent layers in the pavement.

Moving load deflection

This instrument, known as the *deflectograph*, travels at a speed of about 3 km/h and is able to measure the deflection at 3 metre intervals in both wheel tracks. With modification, it is possible to obtain the 'deflection bowl', thus enabling back-analysis of the results in a similar manner to the falling weight deflectometer.

Steady-state vibratory load deflection

This class of instrument also works from a stationary position and applies a small alternating force of about 2 kN to the pavement surface, typically at a frequency of 8 Hz. The deflection induced by this is measured.

Pavement texture and friction

Machines are available for the high-speed recording of texture depth, and of two types of skidding resistance measurement: sideways force and braking force.

Macro-texture or texture depth

These devices use technology developed from the laser profilometers to record the macro-texture. They are available as a trailer that can be towed at road speed, and as a hand-operated device for use in less extensive surveys.

Sideways force

These devices are driven at a known and constant speed (usually 50 km/h) with a test wheel off-set at 20° to the line of travel. The wheel has a known constant load applied to it, and the sideways force along the axis of rotation of the wheel is measured by means of a load transducer. The ratio of these two forces is the friction coefficient, known as the *sideways force coefficient* (SFC).

Braking force

These devices were originally developed for measuring the skidding properties of airfield pavements. They consist of a single wheel trailer which can be towed behind a vehicle that is spraying a track of water on the road. The wheel on the trailer can be braked until it locks to give a reading of the *braking force coefficient*.

Characteristics of methods and sampling rates

Some features of the different methods available for pavement assessment are summarised in Table 5.12, and the characteristic options for deflection methods are given in Table 5.13. Sampling rates for pavement assessment are given in Table 5.14.

Table 5.12 *Methods of measurement for pavement assessment*

	Summary of measurement method	Alternative methods of measurement
IQL-I	Detailed data for individual layers	(1) NDT + (LR or AL) (2) FS + LS + (LL or CR)
IQL-II	Summary data for individual layers	(1) NDT + (FS or CR) (2) FS + (LS or LC) (3) FS + LC
IQL-III	Summary pavement strength index	(1) NDT (2) FS + FC (3) CR
IQL-IV	Category of (relative) pavement strength	(1) Visual survey (2) Inferences from condition data

Notes:
- AL   Material loading behaviour algorithms
- CR   Construction records
- FC   Field material classification
- FS   Field shear test (e.g. cone penetrometer)
- LC   Laboratory field classification
- LL   Laboratory material loading behaviour
- LS   Laboratory material stiffness test
- NDT   Non-destructive test (deflection)

Source: Paterson (1991).

5.7 Data management

5.7.1 Computer-based data management

Effective management requires appropriate information to support management decisions, and the quantitative basis for this is provided through data. The processing of data is facilitated by the use of computer-based systems. Care must be taken in the choice of the most appropriate system. It is recommended that a systematic approach is taken to this, involving steps of identifying objectives, users and outputs of systems, in order to identify the requirement for data and models, before finally selecting appropriate software and hardware. Systems need to be structured to support the primary management functions of: planning, programming, preparation and operations.

Table 5.13 *Characteristics of options for deflection measurement*

Measurement method	Device type	Operating principle	Applied load (kN)	Capital cost	Operating cost[1]	Collection speed (min per reading)
Moving wheel loads	Benkelman beam	Manual + truck	80	H	D	10
	Travelling deflectometer	Automated vehicle	80	B	D	0.6
	Deflectograph	Automated vehicle	100–130	C	D	0.6
Steady-state vibratory	Dynaflect	8 Hz trailer-mounted	0.5	E	E	1.3
	Road rater	6–60 Hz trailer-mounted	0.1–24	E	E	2.7
Impulse load	Falling weight deflectometer	Impact trailer-mounted	15–270[2]	D	E	1.3

Notes:
B US$200 000–400 000
C US$100 000–200 000
D US$60 000–100 000
E US$30 000–60 000
H US$<2000
(1) Annual cost for regular usage
(2) Range varies with make.
Source: Paterson (1991).

It is convenient to differentiate between *information systems* and *decision-support systems* (Paterson and Scullion, 1990).

Road information systems

These collect, organise and store data about the road network, and provide facilities for reports to be produced on these data, in a variety of formats. Information systems have no ability to process these data, although, in order to produce reports, selections from among different types of data and summaries of data may be used. A normal feature of these systems is the incorporation of some kind of database. For example, a road information system may store details about all the sections of road that make up a network; it might be possible to produce reports on this showing all the sections of road in one geographic district, or all the gravel roads in the network and their total length, or all the paved roads in a particular district carrying more than 1000 vehicles per day. The actual reports available will depend on the information stored, the way that it is stored, and the capabilities of the reporting functions of the system (see Figure 5.4).

Table 5.14 *Recommended minimum spatial sampling rates and methods for pavement assessment*

Management function	Minimum sampling rate/interval	Information quality level	Viable assessment methods
Planning	2–5% network length (stratified random sample) 0.5 km per 10 km	IQL-IV or IQL-III	Defl based on FWD or BB; SN based on CR or DCP; Resid based on visual survey
Programming	0.3–1.0 km (min. 5 points per section) JRP: at slab centre and joints	IQL-III or IQL-II	Defl based on FWD or DG
Preparation (project design)	20–200 m (min. 5 points per section) in each wheel path RP: 20–40 m in outer wheel path	Minor roads: IQL-III or IQL-II Major roads: IQL-II or IQL-I	Defl based on FWD, DG or BB; SN based on DCP; plus MS where necessary
Operations, research and special investigation	3–20 m intervals RP: every joint or crack	IQL-II or IQL-I	Defl based on NDT plus MS

Notes:
BB Benkelman beam
CR Construction records of layer thickness and strength
DCP Dynamic cone penetrometer
Defl Deflection measurements at surface
DG Mobile deflectograph
FWD Falling weight deflectometer
JRP Jointed rigid pavements
MS Field or laboratory materials tests, as required
Resid Residual life
RP Rigid pavements
SN Structural number computation

Source: Paterson (1991).

Road decision-support systems

These differ from information systems in having the ability to *process* input data, using functions or algorithms, before producing reports; such processing enables analysis to be carried out which can guide and assist users making management decisions about the network. Examples might include: the identification of the total length of side drainage ditches in the network and the application of a unit cost to this to enable the annual ditch clearing cost to be estimated; the comparison of the severity and extent of defects on individual road sections with threshold values which indicate whether maintenance treatments are required; the determination of the way that budgets should be spent, using pre-defined rules, when the total cost of maintenance treatments identified exceeds the total amount of money that is likely to be available. Normally, a *decision-support* system will incorporate an *information system*. Decision support systems are sometimes referred to as applications systems (see Figure 5.5).

Note that use of the terms like *maintenance management system* and *pavement management system* can cause confusion, because systems produced by different vendors can often have quite different characteristics, even though they may be referred to by the same type. Some commonly used terminology for different types of system is given in Table 5.15, which suggests how these are likely to relate to the management functions of *planning, programming, preparation* and *operations*. To avoid confusion, the following terms are used in this book in relation to road maintenance management:

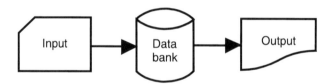

Figure 5.4 *Information system*

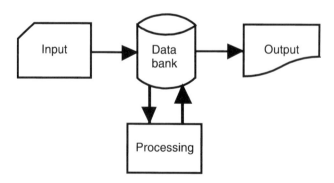

Figure 5.5 *Decision-support system*

- *Information system*
  - network information system.
- *Decision-support system*
  - planning system
  - programming system
  - preparation system
  - operations system.

As the management functions move from planning through to operations, and where computer systems are used to support management activities, computer processes change from being automatic to being undertaken by the manager working interactively with the computer.

5.7.2 System selection and procurement

It is clear from the above that there are many different types of system, and care must be taken to ensure that the chosen one is fit for the purpose for which it is intended, and that its procurement and use will be cost-effective for the organisation (Robinson and May, 1997).

The decision to implement a computerised road management system should be seen, essentially, as a business decision. It should be driven by an assessment of its costs and benefits. But it is not always easy to quantify and achieve benefits, and

Table 5.15 *Examples of relationship of common system descriptions to the management functions*

Management function	Examples of common descriptions
Planning	Strategic analysis system Network planning system Pavement management system
Programming	Programme analysis system Pavement management system Budgeting system
Preparation	Project analysis system Pavement management system Bridge management system Pavement/overlay design system Contract procurement system
Operations	Project management system Maintenance management system Equipment management system Financial management/accounting system

experience suggests that there are real dangers of incurring excessive implementation and recurrent operation costs. It was often thought that benefits would come from replacing people by machines, but experience suggests that this is rarely the case. Often, not only are more staff required, but the required staff need skills that have not traditionally been found within road administrations. Benefits are more likely to come from:

- Improved asset management
- Improved contract and cost control
- Better management information.

Costs will include those for:

- Data collection and updating
- Hardware and software
- Staff training and retraining.

To ensure that a computerised system is an effective solution to a management problem, and not a problem in its own right, the specification and procurement need to be handled carefully. Different approaches to developing road management systems are described by the Transportation Research Board (1994). In the past, road administrations have often reviewed existing commercial products and simply chosen the one which best seemed to fit the current procurement budget. This approach has led to many disappointments. Institutional constraints to implementing systems have been described by several authors (Smith and Hall, 1994; Amekudzi and Attoh-Akine 1996; Robinson and May, 1997). It is far better to consider the system procurement process as a five-phase project in its own right (Britton, 1991, 1994; Smith, R. E., 1994):

1. *Commitment* phase (decide to proceed)
   – obtain commitment from key individuals in the road administration to the system implementation process.
2. *Requirements* phase (decide what is wanted)
   – agree the objective for the system and determine what components the system needs to contain; decisions should be supported by cost–benefit analysis.
3. *Specification* phase (decide what is needed)
   – identify users of the system and the outputs that they will require to support them in their management functions
   – identify data needs and models required to produce these outputs.
4. *Procurement* phase (choose the best solution)
   – identify appropriate software, together with hardware and operating system requirements necessary to support it.
5. *Operations* phase (make the system work)
   – implement the chosen system
   – initial and on-going training
   – managing operation of the system.

This may seem excessive for a potentially small system. However, system implementation costs are usually dominated by annual data collection costs, which are typically five to ten times those of system procurement. If there is a risk that the chosen system is inappropriate, then there would be significant wasted costs in collecting the similarly inappropriate data.

The need for commitment is self-evident, and is not discussed further. Some aspects of the other phases are discussed in turn in the following sub-sections.

5.7.3 Requirements

The identification of system type can be made by considering the policy and objectives of the organisation implementing the system: these aspects are discussed in Chapter 2. Systems should be selected to assist the organisation in meeting its objectives. For example, a *planning* system would be needed to meet an objective concerned with minimising the longer term road user and road maintenance costs in a constrained budget situation; an *operations* system would be needed to provide good costing procedures for works where providing 'value for money' was a stated objective of the organisation.

Compliance with policies and objectives may require that several different types of system are needed. The order in which different types of system are introduced will depend on the details of the local situation and the procedures that are currently operated, as well as the policy and objectives. In such cases, the existing institutional situation is likely to suggest that this process is undertaken cautiously, over a long time scale, using carefully chosen technical assistance. Experience has shown that the temptation to go for a 'big bang' approach should be resisted, and a step-wise implementation should be adopted, ensuring that each step is institutionalised fully before moving on to the next. These steps should form part of a longer-term development plan, where near-term objectives are defined in considerable detail, and longer-term objectives retain flexibility for change and adaptation to allow experience gained from earlier stages to be incorporated.

5.7.4 Specification

It must always be remembered that systems will be used by people to store and process the data that they need to undertake their management functions in the organisation. There is a need therefore, at an early stage, to identify who the potential users of the system will be. It is difficult to make definitive statements about this aspect but, in general, potential users of systems might be characterised by those identified in Table 1.4 as being mainly responsible for undertaking the management functions of *planning, programming, preparation* and *operations*.

The link between the users and the management information or data is the outputs that are produced by the system. These provide the basis of reports that assist users in undertaking their work and, as such, contribute to meeting the road administration's policy. The outputs required will depend on the particular objectives of the administration, on its institutional arrangements, and on the detailed use to which systems will be put. It is not, therefore, possible to be prescriptive in this area. However, in order to provide some guidance, *examples* of the type of outputs that might be required from network information systems and the four types of road management systems are given in Table 5.16.

Table 5.16 *Typical outputs from network information and road management systems*

System type	Typical outputs
Network information systems	• Gazetteer of road sections, in user-defined order, giving attributes of sections (i.e. label, description, other data) • Lists of sections based on user-defined selections of section attributes (e.g. all sections in one geographic district, all gravel roads in the network and their total length, all paved roads in a particular district carrying more than 1000 vehicles per day – the ability to report will depend on both the attributes and the way that they are stored)
Planning systems	• Projected annual capital and recurrent budget requirements to meet road administration standards for a user-defined period into the future • Projected road conditions resulting from the application of pre-defined annual budgets for a user-defined period into the future • Projected road administration costs and road user costs for pre-defined road administration standards, or annual budgets, for a user-defined period into the future • Incremental NPV of adopting one set of standards compared with another, or of adopting one particular stream of annual budgets compared with another
Programming systems	• List of sections, showing recommended treatments and costs that can be funded in the budget year under pre-defined capital and recurrent budget constraint, in priority order and in section order (it should be possible for the user to work interactively with these lists to amend treatments, costs and relative position of projects in the priority list) • List of user-selected sections showing conditions and recommended treatments, in section order • List of user-selected sections showing traffic, axle loading and road user costs, in section order • Projected rolling programme of work over a three year period, which should reflect any user modifications to the list of sections to be treated, as above

Preparation systems	• Project formulation (produced by user working interactively with list of sections requiring treatment, above, possibly by interfacing with design packages, to combine or modify works to produce projects for funding) • Works order for project, including bills of quantities
Operations systems	• Performance standards for works, based on defined activities, plant and equipment costings, materials cost rates, and labour schedule and rates • Work instruction/accomplishment • Weekly labour time summary by person and budget head • Weekly cost summary by activity and budget head, with totals • Annual cost summary by road section, activity and budget head, with totals

The key format for outputs will be tables, and all outputs should be available in this style of presentation. Outputs from *planning* systems are targeted at senior management, and the nature of some of these outputs makes them particularly suitable for presentation as *business graphics* in the form of line graphs, histograms and pie charts. Similarly, many of the outputs from *programming* systems lend themselves to reproduction as *strip diagrams*, which provide a schematic representation of a length of road.

Appropriate data and models will also need to be defined at this stage. Criteria of relevance, appropriateness, reliability and affordability, discussed earlier in this chapter are relevant to this, as is the concept of information quality levels. Models for treatment selection and prioritisation are discussed in Chapters 6 and 7. During the *specification* phase, the investment should be justified in terms of costs and benefits, and a specification written that focuses on the minimum facilities that are needed to deliver those benefits.

5.7.5 Procurement

Having specified the system and, in parallel, justified the investment, it is possible to proceed to the *procurement* phase. The market can be tested for available products, either as off-the-shelf solutions, or as a basis for local customisation, and then compared with the cost of bespoke development. Each case will be slightly different, and engineers are well advised at this stage to seek the advice of IT experts to write operational requirements (i.e. the functional user requirement of the system), detailed specification and contract documents as appropriate. The procurement should be handled by someone who is familiar with computer system projects, but always with reference back to the user in the event of a technical query.

Software

Before discussing software requirements, it should be noted that road management systems do not necessarily require the use of computer software and hardware. There will be examples, particularly on small networks, where manual methods provide the most cost-effective approach. Two example of these, for roads and bridges respectively, are available from the Transport Research Laboratory (TRRL Overseas Unit 1987, 1988b). However, as the volumes of data become larger, computer-based storage and processing becomes essential. The purpose of computer-based systems is to support individuals involved in the management process and not *vice versa*. It is important to keep this in mind at all times and at all stages.

Normally, the most efficient and flexible structure for road management system software is one which is modularised, with integration achieved through a common data bank. This modular structure must reflect the manual operation of the road management process when broken down into functions and tasks. Many proprietary systems lack this modularity and are only available as a complete system, with a resulting loss in flexibility and ability to match the physical management structure.

With modular software, the road *information system*, or data bank, provides the back-bone of the road management system. This comprises the network referencing system around which is built an inventory of the road network. This provides the framework within which all information about, or associated with, the road network is stored and retrieved. The software for this must be flexible enough to accommodate future changes and growth. The modular framework is illustrated in Figure 5.6, which includes typical data items that will be needed, and outputs from the five types of *information* and *decision-support* systems identified earlier. Although such an integrated approach to software development should be a long-term target, in the medium-term most road management systems may only contain part of that shown in the figure.

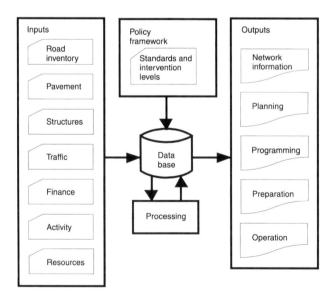

Figure 5.6 *Modular system design*

In computer software terms, the above concept is implemented through a series of application programs operating in conjunction with a database. Application programs will be needed for input, output and processing (models). The main advantage of this system architecture is that it can be tailored to match the management needs and organisational structure of the road administration. All management information required for decision-making is held centrally, while data and technical processing may be decentralised. The division into modules allows the flow of information between modules to be controlled, so that data are checked to ensure quality and consistency before being used by other modules.

This approach represents an ideal situation and does have long-term benefits. Different parts of the system can be developed independently, at different times depending on the resources available, using different software products. The functionality of several of the application modules could be provided by proprietary software, with minimal requirement for customisation. For example, a proprietary *project management* package could be used for work scheduling (Field and Layton, 1995), with a standard database providing the data storage.

The disadvantage of this approach is that long-term considerations will be dictating short-term actions, with the result that the initial solution is more expensive and complicated than a dedicated application. Remembering that the most expensive component of the road management system is the *data*, the key issue is to have the potential for upgrading the system in the future and still to be able to utilise data collected in the past. There will be many benefits when introducing systems for the first time to adopt a very *simple* approach. As operation and use of the simple system becomes institutionalised, and as technology advances, the system can be replaced. Provided that the original system utilises a database or a spreadsheet to store data, it is a relatively straightforward exercise to down-load the data from the original system and to load it up into the new one, although computer system skills are needed to do this. Such an approach may be more in keeping with institutional development requirements while, at the same time, protecting the administration's investment in data.

Hardware

The final decision that should be made when planning the implementation of a road management system is the choice of hardware on which the system will run. The choice of hardware to be used will be influenced by the operating system selected. The basic decision is whether single-user or multi-user access to the management system is required. The decision on the operating system will depend ultimately on the availability of trained personnel capable of maintaining the system: these could be from the road administration itself, or technical support could be procured from local computing companies.

Once the requirements of the road management system software have been defined, and the choice of operating system has been made, the choice of hardware will, in most cases, be self-evident. A system based on microcomputers should be considered as a starting point in many cases because of the availability of appropriate and back-up skills and experience. But the use of work stations should not be overlooked, particularly where large data volumes are anticipated.

5.7.6 Implementation

By the time the *implementation* phase starts, expectations will probably be rising among potential users, and experience shows that this can often be the time of greatest disappointment. Unrealistic delivery schedules, and underestimates by the road administration of the time required to collect, load and validate data into the system can all contribute to delays and cost escalation at this stage. Firm project management will be required.

Where the institutional appraisal and the business case suggest that the introduction of systems is both appropriate and timely, then the process of their implementation should take note of the conclusions from a study (Robinson, 1988) which noted that for successful implementation, systems require:

- Agreement on the aims and methods of implementation between the client and the consultant.
- Sufficiently motivated staff, properly supervised.
- An integrated training and retraining programme.

Human resource issues of staffing and training have already been discussed in Chapter 2. Experience suggests that staffing and training needs in connection with implementing road information and management systems are usually underestimated considerably. The success of any implementation will depend crucially on user involvement, in all aspects of the decision-making process, and on well-trained staff to implement the final system.

It may be seen that the types of management system are many and varied. However, it may be helpful to note the advice given in Box 5.2 when implementing systems.

Box 5.2   *The Danish Road Laboratory's 'ten commandments' for implementing systems*

1. Believe that it is worth doing
2. Only start if resources will be allocated
3. Get acceptance within the organisation
4. Give prestige to those participating
5. Make reasonable, but indispensable, demands
6. Develop step by step
7. Implementation must be staged
8. Each stage should improve current management
9. Make investments in education and training
10. Invest in data-updating starting yesterday.

Source: Schacke and Ertman Larsen (1990).

6 Treatment Selection

- Scheduled and condition-responsive approaches
- Common features
  - methods of defect assessment
  - precedence rules
  - sectioning
- Defect-based rules
- Condition index-based rules
- Complex rules
  - non-transparent methods
  - expert systems
- Optimisation approach
- Treatment selection methods for gravel roads
  - defects and deterioration
  - treatment selection and design
  - assessment of need
- Contract packaging

6.1 Scheduled and condition-responsive methods

6.1.1 Approaches to treatment selection

The use of standard rules for treatment selection ensures that a consistent approach is taken to planning and specifying works throughout the road administration. This helps to ensure that funds are spent to greatest effect, and that each road and part of the network receives its fair share of the budget. The rules should reflect the standards and intervention levels defined in the policy framework.

This chapter discusses the classes of rules that are available for paved and gravel roads. The available rules differ in their degree of complexity. Two fundamentally different types of rules are available:

- *Scheduled*
  Fixed amounts of work (such as a quantity of work in m^2/km) are specified per unit time period (such as one year), or work is specified to be undertaken at fixed intervals of time.
- *Condition-responsive*
  Work is triggered when condition reaches a critical threshold, known as an 'intervention level'.

6.1.2 Scheduled

This type of rule is often used where need is related to environmental conditions (such as cutting back vegetation growth, or cleaning culverts). The approach is also particularly relevant where the deterioration rate is stable. Works can be scheduled at the programming stage when the frequency of intervention is the same year-after-year. Also, where rates of deterioration are rapid, such as for the surface of gravel roads, it is impracticable to respond to defects assessed as the result of condition surveys. Scheduled maintenance may, therefore, be more appropriate.

Routine maintenance of a cyclic nature is normally carried out on a scheduled basis. The schedule for undertaking cyclic activities will often depend on road class to reflect 'level of service' considerations, although this is perhaps hard to justify on purely engineering grounds. An example of such a schedule is shown in Table 6.1. With methods such as this, the frequencies of activities are user-specified, and the values given in the table are examples only. Periodic works, such as surface dressing, may also often be specified this way, particularly where the effect of a severe environment outweighs the damaging effect of traffic.

Table 6.1 *Routine treatment frequencies*

Code	Description	Maintenance standard (no times/year) for road class					
		A	B	C	D	E	F
01	Grass cutting by machine	4	4	2	2	2	2
02	Grass cutting by hand	4	4	2	2	2	2
03	Machine cleaning of V-shaped side drains	2	2	2	2	2	2
04	Machine cleaning of U-shaped side drains	2	2	2	2	2	2
05	Manual cleaning side drains	2	2	2	2	2	2
06	Cleaning culverts	1	1	1	1	1	1

Source: High-Point Rendel.

The treatment selection issues concerned with scheduled methods are related mainly to engineering decisions about the treatment method rather than to fundamental decisions about which treatment to apply. For example, decisions may need to be taken about whether ditches should be cleared using machinery or by hand; this decision may depend on the shape of the ditches, whether or not they are lined, the availability of the appropriate equipment for their maintenance, the relative costs of equipment and labour, the availability of labour, plus other issues such as the political desirability of using labour.

The other area of decision is the frequency at which the work is scheduled. Unfortunately, few models exist which can be used to determine appropriate frequencies for routine maintenance activities related to paved roads. In such cases, decisions must continue to be made on the basis of engineering judgement, based on local observation, and/or political decisions relating to amenity value. Such decisions will also be tempered by the availability of funds for such works.

Many road administrations find it convenient to plan and manage works on a scheduled basis, and the approach is being used increasingly for a wide range of defects and works, except on the most heavily-trafficked roads. This approach is widely used for the planning and programming of works.

6.1.3 Condition-responsive methods

Whereas scheduled rules are relatively easy to specify and implement, there are several methods of specifying condition-responsive rules. The remaining part of this chapter is devoted largely to describing these.

Condition-responsive rules involve the use of intervention levels which, when exceeded, trigger different treatments. It was noted in Chapter 5 that the following indicators of pavement condition should be considered:

- Roughness
- Surface distress
- Structural capacity
- Pavement texture and friction.

OECD (1995) suggests that, in addition to identifying candidate treatment options, decision rules should normally aim to:

- Supply results that are fairly close to customary methods, giving similar results to those based on the judgement of the maintenance engineer.
- Offer technical improvements and increased productivity.
- Grant managers the possibility of applying their own expertise and know-how in order to correct errors and excesses which result inevitably from 'black box' systems.

6.1.4 Methods available

As the decision-making process moves from planning to operations, and from network to project level, the treatment selection methods tend to involve the use of a larger number of indicators and the use of more detailed rules. The following general types of method are available for condition-responsive treatment selection:

- Defect-based rules
- Condition index-based rules
- Complex rules.

In addition, a fourth class of method is available that uses no rules. Instead, an 'optimisation' approach is used for treatment selection based on considerations only of treatment cost and life.

Treatment selection rules are discussed under these general headings for paved roads in the following sections. However, the division into these three classes is relatively arbitrary, and is introduced only to facilitate description of the different methods. In practice, a continuum of methods exist, varying from the very simple to the extremely complex. Furthermore, some methods do not lend themselves to easy classification. Appropriate rules for gravel roads also require special consideration, and these are discussed separately. However, there are certain common features of treatment selection methods, and these are discussed next.

6.2 Common features

6.2.1 Methods of defect assessment

It will be seen that each of the methods described requires physical inspection to determine the severity and extent of defectiveness. It is recommended that objective measurement of defects should be used in all possible cases. The nature of defects means that, even with measurements, problems of repeatability and reproducibility will still arise. Problems are likely to be considerably greater when defects are rated rather than measured. There is a need to 'calibrate' the people doing the inspections, particularly in different regions where the perception of problems may differ. In general, machine-based methods are preferred to manual measurement.

There will often be advantage when using defect assessment for monitoring the appropriateness of maintenance objectives and standards, in being able to utilise actual measurements of condition, rather than ratings which already contain an element of 'political' decision. With increased use of maintenance by contract (see Chapter 9), it becomes increasingly important to be able to specify items in quantifiable terms for contractual purposes.

However, it should also be recognised that in some situations, such as with the recording of off-carriageway defects, measurements may be both difficult and time-consuming, if they are possible at all. In such cases, it may be preferable to schedule works over time, wherever possible, or to specify treatments and quantities directly on the basis of engineering judgement.

6.2.2 Precedence rules

Most treatment selection methods that are rules-based make use of precedence rules. Precedence rules identify which treatment should be applied when the need for more than one treatment is identified. This appears to be trivial but, when using computer-based methods, such rules must be specified to replace the intuitive knowledge of the maintenance engineer. For example, if the level of surface distress is sufficient to trigger resealing, and an overlay is needed to address a structural defect, then the overlay would be applied rather than the resealing; this is because the overlay would correct both the surface and structural problems.

A simple treatment precedence used at network level could be based on works types, as defined in Chapter 1:

1. Pavement reconstruction
2. Overlay

3. Resurfacing
4. Preventive
5. Reactive.

At project level, precedence rules might be couched in terms of activities, again using definitions given in Chapter 1:

1. Full pavement reconstruction
2. Mill and replace
3. Inlay
4. Dense graded asphalt overlay
5. Thin overlay
6. Single surface dressing
7. Fog seal/surface rejuvenation
8. Patching
9. Crack sealing.

All of the methods discussed in the following sections have either an explicit or implicit precedence rules associated with them.

6.2.3 *Sectioning*

Treatment selection rules can be applied either to lengths of road that are fixed or of variable length. This issue was referred to in sub-section 5.4.2. Fixed length sectioning means that treatments are always applied to pre-defined lengths of road: often these are the sections or sub-sections used to reference the network. A more sophisticated and flexible method makes use of the concept of 'dynamic sectioning', or 'dynamic segmentation'.

Different defects are assessed along the length of the road and, for each, the chainage is recorded at which the level of defectiveness changes. In this way, a section of road is broken up into lengths of road with uniform defectiveness, sometimes known as 'defect lengths', as shown in Figure 6.1. Treatment selection rules can then be applied to these individual 'defect lengths' to select the most appropriate treatment for the range of defects that are present. In theory, such a method is efficient since treatments are only applied over the lengths for which they are warranted, and not over a whole section, only part of which may be defective.

However, in practice, defect lengths resulting from this method can be quite short, with the result that it is sometimes impracticable or uneconomic to apply different treatments to each. Algorithms are usually used to combine short defect lengths into longer sections of road of a minimum length to enable appropriate treatments to be applied in a cost-effective manner. These 'treatment lengths', although sub-optimal in terms of the length over which a common treatment should be applied, will still normally provide a more cost-effective solution than applying the treatment over the whole section.

This method of applying treatments is normally found in association with the more sophisticated treatment selection methods.

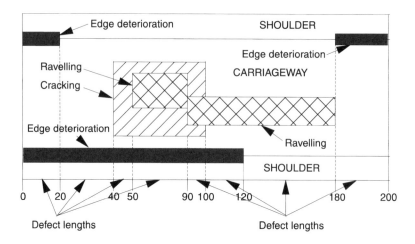

Figure 6.1 *The determination of defect lengths by dynamic sectioning*

6.3 Defect-based rules

These methods make use of a matrix of relationships between measured or assessed defects and treatments. A particular treatment is triggered if one or more defects exceed their respective intervention levels. Relationships may contain logical 'AND' or 'OR' functions to combine defects. The method is conceptualised in Figure 6.2 for a two-dimensional relationship of defects using 'AND' relationships. The figure also shows how different sets of rules may be required for different functional hierarchies to reflect the requirements of the policy framework.

An example of a simple rules method of treatment selection is that from *Overseas Road Note 1* (TRRL Overseas Unit, 1987). The table of intervention levels for paved roads is reproduced in Table 6.2. In this particular example, the recommendations are limited to treatments that can be identified from visual inspections only; where more substantial treatments are needed, the method recommends that 'further investigation' is undertaken.

Examples of a graphical approach to treatment selection are shown in Figures 6.3 and 6.4 (Phillips, 1994). An example of a method of determining treatments for off-carriageway features is shown in Table 6.3. Note that, with off-carriageway features, the difficulty of measuring defects results in intervention criteria tending to be more subjective than those that are used for carriageway defects.

6.4 Rules based on condition indices

Condition indices combine defects into groups using functional relationships for treatment selection purposes. They provide:

- A convenient grouping of defects as an interim step in a calculation or algorithm for determining treatments.
- A generic statement about defectiveness of the four main defect groups of roughness, surface distress, structural capacity, and pavement texture and friction.

	Hierarchy 1	
Thresholds for defect a	Thresholds for defect b	Treatment
$a_1 > t_1$	$b_{11} > t_{11}$	x_a
	$b_{12} > t_{12}$	x_b
	$b_{13} > t_{13}$	x_c
$a_2 > t_2$	$b_{21} > t_{21}$	x_d
	$b_{22} > t_{22}$	x_e
$a_3 > t_3$	$b_{31} > t_{31}$	x_f

(Hierarchy 2, Hierarchy 3 shown as layered panels behind Hierarchy 1)

Figure 6.2 *Format for treatment selection using defect-based rules*

Such an approach is helpful, since most treatments can be seen to be correcting one or more of the four defect groups (Phillips, 1994; Duffell and Pan, 1996). For example:

- A surface distress condition index might be a function of the defects of surface cracking, fretting and bleeding.
- A structural adequacy condition index might be a function of the defects of structural cracking, rutting and deflection.

Condition indices may also be used to group defects for the road edge or for off-carriageway features.

Approaches to treatment selection embodied within these methods are used mostly at the *programming* stage to identify recommended treatments. Many systems then allow the engineer to interact with these recommendations to take account of local knowledge and to modify or refine the treatment choice. An example of a method that uses condition indices is given in Box 6.1. Further examples are given by Jackson *et al.*, (1996). It will be seen in this method that, in fact, a combination of 'defect-based' and 'condition index-based' approaches is used, and this is common in many practical systems. At the *preparation* stage, detailed engineering investigations are normally undertaken for those treatment choices involving significant cost to refine and design the detailed treatment choice.

Table 6.2 Table of intervention levels for paved roads

Defect	Level	Extent (% of sub-section length)	Climate/ traffic category	Defect	Extent (% of sub-section length)	Action
Stripping or fretting	Any	< 10	All	—		Local sealing
		> 20	All	—		Surface dress
Fatting-up or bleeding	—	—	All	—		No action
Pot-holes	Any	—	All	—		Patch
Edge damage	Erosion from original edge > 150 mm	> 20	All	—		Patch road edge and repair shoulder
Edge step	> 50 mm	> 50	All	—		Reconstuct shoulder
Wheel track rutting (surface dressing on granular base)	< 10 mm	—	Rainfall > 1500 mm/yr OR Traffic >1000 vpd	Wheel track cracking	< 5	Seal cracks
					> 5	Surface dress
				Non-wheel track cracking	< 10	Seal cracks
					> 10	Surface dress
			Rainfall < 1500 mm/yr AND Traffic < 1000 vpd	Wheel track cracking	< 10	Seal cracks
					> 10	Surface dress
				Non-wheel track cracking	< 20	None
					> 20	Surface dress
	10–15 mm	> 10	All	Any cracking	—	Treat cracks depending on extent as above if rate of change of rut depth is slow
						Further investigation if rate of change of rut depth is fast
	>15 mm	<10	All	Cracking only associated with local ruts	—	Patch
				Other cracking	—	Patch excess rutting and treat cracks depending on extent as above
		> 10	All	Any cracking	—	Further investigation

Defect	Level	Extent (% of sub-section length)	Climate/ traffic category	Defect	Extent (% of sub-section length)	Action
Wheel track rutting (asphaltic concrete granular on base)	< 10 mm	—	Rainfall > 1500 mm/yr OR Traffic > 1000 vpd	Any cracking	< 5	Seal cracks
					5–10	Surface dress
					> 10	Further investigation
			Rainfall < 1500 mm/yr AND Traffic < 1000 vpd	Any cracking	< 10	Seal cracks
					10–20	Surface dress
					> 20	Further investigation
	> 10 mm	< 5	All	Cracking only associated with local ruts	—	Patch
				Other cracking	—	Patch excess rutting and treat cracks depending on extent as above
		> 5	All	Any cracking	—	Treat cracks depending on extent as rate of above if change of rut depth is slow
						Further investigation if rate of change of rut depth is fast
Wheel track rutting (asphaltic concrete or surface dressing on stabilised road base)	< 5 mm	—	Rainfall > 1500 mm/yr OR Traffic > 1000 vpd	Any cracking	< 10	Seal cracks
					> 10	Seal cracks and surface dress
			Rainfall < 1500 mm OR Traffic > 1000 vpd	Any cracking	< 20	Seal cracks
					> 20	Seal cracks and surface dress
	5–10 mm	> 10	All	Any cracking	—	Treat cracks depending on extent as above if rate of change of rut depth is slow
						Further investigation if rate of change of rut depth is fast
	> 10 mm	< 5	All	Cracking only associated with local ruts	—	Patch
				Other cracking	—	Patch excess rutting and treat cracks depending on extent as above
		> 5	All	Any cracking	—	Further investigation

Source: TRRL Overseas Unit (1987). Crown copyright 1987.
Reproduced by permission of the Controller of H. M. Stationery Office.

Road Maintenance Management

Table 6.3 *Intervention levels for off-carriageway features*

Treatment	Intervention level
Kerb replacement/repair	Missing or damaged kerbs that need repairing or replacement (engineering judgement)
Grass cutting	Height of vegetation is such that it interferes with the line of sight
Shoulder repair	Deformation or scour, which is hazardous to traffic, or which is endangering the structure of the road (engineering judgement)
Footway repair	Cracking or settlement with sudden trips of greater than 25 mm
Cleaning side drains	Silted up or blocked to the extent that the free flow of water is impeded
Side drain repair	Scoured or damaged to the extent that repairs are necessary (engineering judgement)
Cleaning culverts	Silted up or blocked to the extent that the free flow of water is impeded
Culvert repair	Scoured or damaged to the extent that repairs are necessary (engineering judgement)
Guard rail repair	Damaged or missing to the extent that they are a hazard to traffic or fail to perform their proper judgement
Retaining wall repair	Damaged to the extent that they are a hazard to traffic, or the structure is in danger of collapse
Removal of landslide debris	Debris on the carriageway

Source: High-Point Rendel.

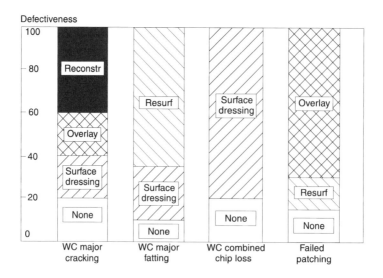

Figure 6.3 *Example of a graphical treatment selection method for single defects* (*Source*: Phillips, 1994)

Treatment Selection 171

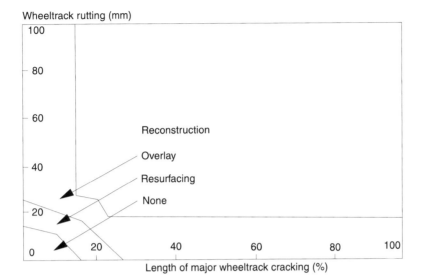

Figure 6.4 *Example of a graphical treatment selection method for pairs of defects*
(*Source:* Phillips, 1994)

| **Box 6.1** | **Example of a condition index-based method** |

The method uses several measures of condition, which are grouped three dimensionally:

- Those which are indicative of surfacing problems only
- Those which are indicative of structural problems
- Roughness

This method of treatment selection recognises that different treatments are needed to correct defects that are either 'surface', 'structural' or 'roughness' in nature; and also that there will often be a combination of these defect types on any section of road. Thus, the method works in four steps.

Step 1
The condition of bituminous pavements is quantified by a *structural condition index* (CI_{Struct}) and a *surface condition index* (CI_{Surf}). These indices are defined as functions of 'structural' and 'surface' defects, as follows:

CI_{Struct} = [A/5], *with a maximum value of 5*
CI_{Surf} = [(C + R)/15], *with a maximum value of 5*

where A = Average rut depth (mm)
       C = Cracking (%)
       R = Ravelling (%).

Step 2
Carriageway treatments are then determined according to the matrix of condition indices, shown below, which shows a treatment for each combination of structural and surface condition index. All of the relationships and parameters in this example are user-definable.

(*continued*)

CI_{Surf}	CI_{Struct}					
	0	1	2	3	4	5
0	—	—	—	—	—	Reconstr
1	—	—	—	—	—	Reconstr
2	Reseal	Reseal	Reseal	Reseal	Reconstr	Reconstr
3	Reseal	Overlay	Overlay	Overlay	Reconstr	Reconstr
4	Overlay	Overlay	Overlay	Overlay	Reconstr	Reconstr
5	Reconstr	Reconstr	Reconstr	Reconstr	Reconstr	Reconstr

Step 3
All pot-holes, on all classes of road, are patched as a reactive maintenance treatment. Where the need for edge repair is identified by the road inspection, its inclusion as a separated treatment is decided in the following way:

- If pavement reconstruction is also identified as being required, then the edge is repaired as part of that reconstruction, and not costed separately
- If an overlay or reseal is identified as being required, then all edge repairs identified are carried out by haunching prior to the overlay or reseal
- If no other periodic treatment is identified, then edge repairs are carried out by edge patching only in association with any carriageway patching.

Step 4
On heavily trafficked roads, where vehicle operating costs are important because of their economic effect, the method also includes the effect of roughness on treatment selection. As a pavement deteriorates, its roughness increases and, at different levels of roughness, different treatments become appropriate. The treatment selection rules based on roughness are given below.

Roughness (m/kmIRI)	Treatment
> 10	Pavement reconstruction
4–10	Overlay
< 4	(No treatment)

Thus, for any pavement condition defined in terms of different defects, alternative treatments will be selected based on the structural and surface condition indices, and on roughness. In order to choose between these, reference is made to the precedence rules shown below. The highest priority treatment from the alternatives based on condition indices and roughness is selected as the recommended treatment.

Precedence	Treatment
1	Pavement reconstruction
2	Overlay
3	Reseal
4	Patching
*	Edge repair

* The precedence of edge repair follows that of other treatments with which it is associated.

Source: High-Point Rendel.

6.5 Complex rules approaches

6.5.1 Non-transparent methods

The methods of treatment selection described above are all simple enough for the method of treatment choice to be transparent to the user. However, as the number of criteria used, and the complexity of the method increases, the degree of understandability diminishes. This is best illustrated by an example based on the treatment selection method used in the UKPMS system (Phillips, 1994). This has been shown to give results that are consistent with engineering expectations, but whose method of operation is difficult to follow manually. The method follows the steps shown in Figure 6.5; the rules applied at each of the steps are described in Box 6.2.

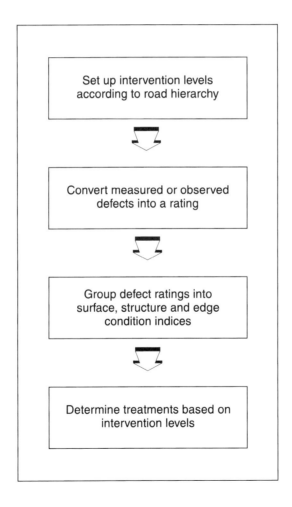

Figure 6.5 *Steps in the UKPMS treatment selection process*

6.5.2 Expert systems

Knowledge based expert systems were originally developed by the medical profession for the diagnosis of multiple symptoms in order to determine the most appropriate treatment. The required approach for road maintenance treatment selection can be seen, in essence, to be similar. The concept is that the collective experience of a large number of professionals is encapsulated in a series of rules, normally specified in the form of complex decision trees. These may be programmed into an expert system software 'shell' so that this collective experience can be utilised easily by others. Expert systems operate by selecting feasible maintenance and rehabilitation options from a set of available alternatives by considering combinations of factors that represent local experience (Ritchie *et al.*, 1986; Wang *et al.*, 1994).

Box 6.2   *Example of a non-transparent treatment selection method*

Treatments are determined on the basis of specified intervention levels (IL) which vary according to road hierarchy and measured condition index. A typical set of intervention levels is shown below.

Intervention level	Primary road	Secondary road	Tertiary road
IL6	90	90	90
IL5	80	80	80
IL4	90	90	90
IL3	70	70	70
IL2	40	40	40
IL1	50	55	60

The method converts each measured or observed defect into a 'rating' on a scale of 1–100, with '1' representing excellent condition. The purpose of this is primarily to enable defects of different types, and on different features to be compared, one with another, on a common scale. For example, rating carriageway and footway defects makes it possible to compare directly the relative severity of defects on these features in terms of which should have the priority for repair. Similarly, comparisons can be made more easily between the severity of defects on bituminous and concrete carriageways. However, the choice of rating rules must be subjective and will normally be made to reflect pre-defined policies about the importance and priority that should be afforded to different features. Examples of rating rules are shown below.

Whole carriageway major cracking		Wheel track major cracking		Rut depth	
Per cent of area	Rating	Per cent of area	Rating	mm	Rating
0	0	0	0	0	0
20	53	15	67	15	67
40	74	25	93	25	93
60	95	100	100	100	100
100	100				

Box 6.2 (*contd.*)

The method also groups defects into those affecting the structure, the surface and the edge, and derives 'condition indices' for each of these. Rules are defined for these indices, such as shown below for structural condition indices of flexible pavements. In this case, the structural condition index is given by the highest rating of:

- 1.0 × residual life (obtained from deflection measurements)
- 0.95 × whole carriageway major cracking
- (0.5 × residual life) + (0.6 × whole carriageway major cracking)
- (0.7 × wheel track major cracking) + (0.3 × wheel track rutting)
- [(0.3 × wheel track major cracking) + (0.7 × wheel track rutting)] or (0.8 × failed patching/reinstatement)

Treatments are then determined on the basis of specified intervention levels and the calculated condition indices, as shown below. Again, the actual values adopted are user-definable.

Structural CI	Surface CI	Edge CI	Treatment
$\geq$ IL6	Any	Any	Reconstruct
$\geq$ IL5	Any	$\geq$ IL4	Reconstruct
$\geq$ IL5	Any	< IL1	Partial reconstruct
$\geq$ IL3	Any	$\geq$ IL4	Edge repair + overlay
$\geq$ IL3	Any	< IL1	Overlay
$\geq$ IL3	Any	$\geq$ IL1	Edge repair + overlay
$\geq$ IL1	Any	$\geq$ IL4	Edge repair + overlay
$\geq$ IL1	Any	$\geq$ IL1	Edge repair + overlay
< IL1	< IL1	$\geq$ IL4	Reconstruct edge
< IL1	< IL1	$\geq$ IL1	Partial reconstruct edge
< IL1	> IL1	$\geq$ IL4	Reconstruct edge + overlay
< IL1	> IL1	$\geq$ IL1	Partial reconstruct edge + overlay
< IL1	> IL1	< IL1	Inlay
< IL1	Any	< IL1	Inlay
< IL1	$\geq$ IL2	< IL1	Surface dress
Any	$\geq$ IL2	$\geq$ IL4	Reconstruct edge + surface dress
Any	$\geq$ IL2	$\geq$ IL1	Partial reconstruct edge + surface dress

Adapted from: Phillips (1994).

It has been argued (Evdorides and Snaith, 1996) that expert systems are needed for pavement assessment because, despite recent developments, the methods currently available are not sufficiently accurate. Reasons for this include:

- The nature of road pavements and their loading characteristics are still not fully understood.

- Current analytical procedures tend to be based either on empirical methods or on overly simplified approaches.
- Current mechanistic procedures of treatment design largely ignore functional road condition data which can give a clue to the cause of deterioration.

Most expert systems are programmed into interactive computer systems. Users are presented with a series of questions and, depending on answers given to these, the subsequent questions will differ. In simple terms, this process can be represented as a tree structure. The user is led from the starting point (the 'root' or 'trunk') to the eventual recommendation. An example of this is shown in Figure 6.6 for alligator and block cracking. A further example of part of a decision tree is shown in Table 6.4.

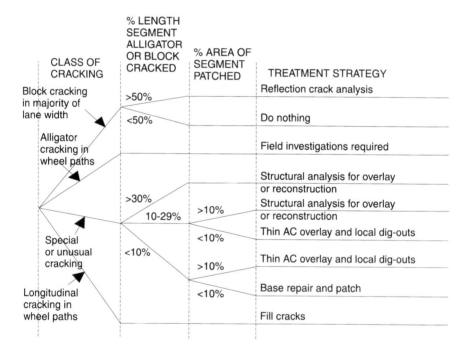

Figure 6.6 *Decision tree for alligator and block cracking* (*Source*: Peterson, 1987)

The general role of expert systems in treatment selection is first to identify the generic or principal treatment option, and then to develop more detailed, but feasible options, within this generic type. It should be noted that although expert systems can capture genuine expert knowledge, there is also a danger that their use can institutionalise past bad practices.

Treatment Selection

Table 6.4 *Matrix form of decision tree for treatment selection*

Distress	Combinations of distress (read vertically)													
PSI < 4.0	N	N	N	N	N	N	N	N	Y	Y	Y	Y	Y	Y
Major cracking	N	N	N	N	Y	Y	Y	Y						
Rutting > 30%	Y	N	N	N										
Ravelling > 30%		Y	N	N										
Bleeding > 30%			Y	N										
Alligator crack > 30%					N	N	N	Y						
Edge crack > 30%					N	N	Y							
Longitudinal crack > 30%					N	Y								
Excess crown									Y	N	N			
AADT > 5000									N	Y		N	Y	N	Y
Alligator crack major											N	N	Y	Y
Feasible rehabilitation options	3 4 6 11	1 5 7 12	1 8 12	2 4 5	2 5 7	3 4 6 9 10	2 6 9 11	3 6 10 11	4 10	1 4 9 11	2 9 4 5 9 10	2 9 4 5 9 10	3 9 11	2 4 6 10	3 9 11

Notes: Y = specified condition is met
N = specified condition is not met
(blank = specified condition is not considered for this treatment)

Rehabilitation options:
  (1) 1" overlay
  (2) 2" overlay
  (3) 3" overlay
  (4) Mill 1" + chip seal
  (5) Recycle + 1" overlay
  (6) Recycle + 2" overlay
  (7) Plane + 1" overlay
  (8) Plane + 2" overlay
  (9) Plane + 3" overlay
  (10) Reconstr 2" AC + 4" base
  (11) Reconstr 2" AC + 6" base
  (12) Chip seal

Source: Haas *et al.* (1994).

6.6 Optimisation approach

This treatment selection method differs from all of the others because it does not use intervention levels. Instead, an optimisation problem is set up that chooses between all possible treatment options applied to all sections of road in the network. Treatment options are characterised by their cost, life, and their impact on road condition. The optimisation method chooses one treatment option for each section in such a way that either:

- The quality of network condition over time is maximised, or
- The road maintenance costs over time is minimised.

Other factors, such as the implications on road user costs, may also be taken into account.

This process is particularly useful at network level, where the plan of programme is subject to budget constraint. This means that there may be insufficient funds to carry out treatments on all sections that are defective. The method chooses the best combination of treatments across sections that meets the relevant optimisation criteria.

The method of solution of the optimisation problem is described in section 7.4.

6.7 Issues arising from condition-responsive methods

The three different methods that include the concept of intervention levels, in one form or another, require recognition of the following issues:

- Logic would indicate that intervention levels need to be different for different levels and loadings of traffic, for different treatment costs and in different environments; the use of models such as HDM-III (Watanatada *et al.*, 1987) can sometimes assist in determining these.
- Different treatments correct different defects or groups of defects.

The first of these points suggests that any method used will require local calibration but that, even when calibrated to reflect different costs and environmental situations, intervention levels will need to be different for different levels of traffic. The difficulty of having continuous relationships between treatment rules, or intervention levels, and traffic levels suggests that a more pragmatic approach is needed, possibly with different sets of rules applying to different bands of traffic flow and loading. Such 'bands' would normally be consistent with the hierarchies discussed in Chapter 1. This approach has been used, for example, in Cyprus (Kerali *et al.*, 1991).

The second of these points suggests that methods should recognise that defects and treatments can be grouped in some way. The approach using condition indices takes this into account.

Current treatment selection models tend not to separate decisions about reactive and periodic treatments but, rather, treat them as a 'continuum' of decision-making. This is reflected in the use of precedence rules described in sub-section 6.2.2.

It is perhaps surprising that there is a paucity of research to support any modelling of the treatment selection process and the selection of thresholds which trigger particular maintenance treatments in the different situations. For example, given a particular combination of defects on a road carrying a given traffic loading in a known environment, what will be the implications of future performance as a result of applying a reseal, of overlays of different thicknesses or strengths, of a full or partial pavement reconstruction, of simply sealing cracks and filling pot-holes, or of doing nothing? Such questions are still difficult to answer.

However, it must be recognised that such questions are actually complex, and solutions would involve the input and modelling of large numbers of parameters. It is for this reason that optimisation approaches are being pursued. These methods rely on predictions of future road deterioration, given knowledge of conditions at the present time and other parameters. They utilise prediction relationships such as those in HDM-III (Watanatada *et al.*, 1987).

The various methods available reflect different ways of visualising complex decision processes. The appropriate choice of method will differ in individual circumstances, depending on the method of defect identification that is considered cost-effective, and on the degree of transparency of methodology that the specific users of the method require. There may also be a requirement for consistency of method to be used at planning, programming and preparation levels, which may affect the choice.

6.8 Treatment selection for gravel roads

6.8.1 Incidence of gravel roads

For most countries outside Western Europe, a significant proportion of the road network consists of gravel and earth roads. For many countries, unpaved roads represent the majority of the road network. Even in the industrialised countries of the world, there are often significant lengths of unpaved roads, particularly in forests and on recreational routes. The management of unpaved roads poses many problems that are different in nature from those of paved roads, so it is appropriate that they should be discussed here. However, this book does not attempt to address the many particular issues concerned with the management of earth roads, where the function is purely to provide access, and on which traffic levels are very low. Other books deal specifically with the design, construction and maintenance problem of these (for example, Hindson, 1983). This section concentrates on treatment selection for gravel roads, where they differ from that of paved roads.

6.8.2 Defects and deterioration

Defects encountered

Defects on gravel roads can be considered under the same general headings as those introduced in Chapter 5, but some modification to these is necessary.

Defects under the heading of *pavement texture and friction* are not normally considered for gravel roads. Although the slipperiness of gravel roads can be a problem, particularly after heavy rain or when there is frost, the traffic levels on these roads do not normally warrant maintenance treatments to correct this defect. Any skid resistance requirements are normally taken into account by the choice of gravels used at the time or construction or during any regravelling operations.

The additional defect of *dust* is also necessary for consideration on gravel roads.

As a result, the defect parameters on gravel roads normally considered are:

- Roughness
- Surface distress
- Structural adequacy
- Dust.

Roughness

Roughness defects on gravel roads are similar in appearance to those on paved roads, although their magnitude is normally much larger. Roads become rough because the material particles making up the road surface are moved around as a result of the effects of traffic and climate. One result of increased road roughness is that vehicle operating costs increase (Watanatada *et al.*,1987). Where road maintenance is not carried out at frequent intervals, the increase in operating costs can be substantial. Relationships have been derived for the increase in roughness as a function of traffic, rainfall and other parameters, and an example of these is given by Paterson (1987).

The problem of corrugations is a particular example of the general case of roughness, and the existence of these causes significant problems both for the vehicle user and for the road manager. There is now wide acceptance of the proposition that corrugations are formed by forced oscillations at the resonant frequency of vehicles' suspension and tyre system (Heath and Robinson, 1980). However, the specification of materials properties that will inhibit the formation of corrugations has proved to be much more elusive (Millard, 1993), although particle size distribution and plasticity are known to be determinants. Corrugations are a particular concern of the road manager because they induce very high levels of road roughness, with consequentially high vehicle operating costs, and because they tend to form and become severe in a very short period following maintenance activities. The cost of maintenance of providing a corrugation-free road on corrugation-susceptible materials can be very high.

Surface distress

Defects of surface distress are similar to those on paved roads, but defects such as cracking are not an issue. 'Loss of camber' would normally be considered as a defect under this heading, although this would be considered a 'construction fault' on paved roads. In addition, 'rutting' would also be considered, whereas this is a 'structural adequacy' defect on paved roads.

The following defects can therefore be included under this heading:

- Loss of camber
- Rutting
- Pot-holes
- Ravelling or loose material.

Structural adequacy

Defects under this heading need to be considered differently from those on paved roads. The gravel surface of a road can be considered as a 'wasting resource' and, as such, the concept of a pavement 'structural design' for gravel surfacings is not really appropriate (Millard, 1993). Normal practice is to provide a minimum thickness of gravel of specified quality to provide a structural layer to protect the subgrade, and then to replenish this as it is worn away. Structural adequacy is therefore related to the minimum thickness of gravel surfacing that is required to perform this function.

Particularly in dry weather conditions, gravel materials degrade as a result of the abrasive action of traffic and the environment:

- The coarse material, which is able to stand high wheel loads, suffers both from ravelling and loss of material to the side of the road.
- The finer material, which is present to hold the coarse material in place, is blown away as dust.

Although some of the material lost in this way can be brought back onto the road through maintenance actions, the overall losses can be considerable. In many countries, supplies of good road gravels have been depleted to the extent that it is sometimes necessary to haul regravelling materials distances in excess of 100 kilometres. The resulting high costs of regravelling in such cases make it economic to pave roads with a bituminous surfacing at relatively low traffic levels.

Predictive relationships for the loss of gravel from road surfaces have been produced, for example, by Jones (1984b) and Paterson (1987). An example of predictive relationships for lateritic gravels is given in Figure 6.7.

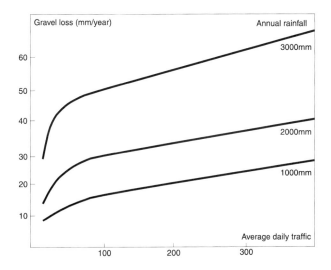

Figure 6.7 *Gravel loss prediction curves for lateritic gravel* (*Source*: Robinson, 1988. Crown copyright 1988. Reproduced by permission of the Controller of H. M. Stationery Office)

Dust

The defect of dust raised by traffic on gravel roads in dry weather is significant because it has an effect both on the population living by the roadside, on the safety and comfort of the road user and on maintenance requirements. In addition to the general unpleasantness and the deposition of dirt and grime, airborne dust can contribute to respiratory and other health problems. In rural areas, deposition of road dust can also decrease crop production adjacent to the road. The reduction in visibility for traffic on the open road can be a contributory factor in road accidents.

Annual losses of material from the road as dust can be significant. Losses of up to 25 tonnes per kilometre per year have been reported by Jones (1984a) for Kenya.

6.8.3 Treatment selection and design

In addition to the needs for road-side maintenance, which are similar to the requirements for paved roads, the following maintenance activities are normally appropriate for gravel roads:

- *Routine*
    - patching/filling
    - grading; heavy grading, light grading or dragging
    - dust control.
- *Periodic (overlay)*
    - regravelling.

Patching is a labour-based activity that can be used to repair pot-holes and other localised surface distress. Arguably, it is only likely to be cost-effective in countries where the cost of labour is cheap; grading is likely to prove more economic in other situations. It is seldom economic and, normally, technically inappropriate to use labour-based methods to repair corrugations or ruts, to replenish lost gravel, or to undertake other activities that require inputs over a wide area (see Chapter 8).

Grading consists of reshaping the existing gravel material on the road surface to provide a smoother ride for vehicles. It can be done, essentially, in three ways. Heavy grading involves scarifying, watering, reshaping and compaction of the gravel material. Light grading and dragging both involve reshaping only, and differ only in the equipment that is used to do this. Light grading utilises motor graders or towed graders, whereas dragging involves the use of 'tolards' or brooms, normally towed behind a tractor (TRRL Overseas Unit, 1985). Obviously heavy grading is more effective and long-lasting than either light grading or dragging, but it is also a more expensive operation. Local studies should normally be carried out to determine the most effective strategy for the use of combinations of these activities within the budget constraints that may apply.

Dust control is normally undertaken by the application of one of the following liquids to the road surface (Millard, 1993):

- Water
- Deliquescent salts, such as calcium chloride
- Organic compounds, including sulphite liquor, molasses, palm and other vegetable oils
- Mineral oils, such as waste fuel oils.

The relief obtained by spraying roads with water is normally very short-lived, particularly in hot climates. Deliquescent salts function by retaining moisture in the surfacing. Organic compounds and mineral oils function by coating and binding the dust particles. Use of such dust palliatives is only economic when they are available as waste materials and, in all cases, their effectiveness is only temporary. Millard (1993) suggests that, when the cost of repeated applications is taken into account, the application of palliatives is likely to be more expensive than a more permanent treatment, such as surface dressing. They may, however, be a useful expedient while awaiting more permanent action.

Regravelling is essentially the same operation as new pavement construction for gravel roads. It is a relatively expensive operation. The treatment of 'spot regravelling' has much in its favour since new material is only applied to those areas of the road where passibility is a problem. This focuses resources in those areas where they are needed most, and this approach can prove to be very cost-effective.

6.8.4 Assessment of need

Patching, dust control and regravelling

In principle, the need for these treatments is determined in the same way as condition-responsive treatments for paved roads described earlier. Normally regravelling works are triggered when gravel thickness is reduced to an unacceptable level, perhaps when the subgrade starts to show through the surfacing material. However, it may prove to be more appropriate to treat patching as a *scheduled* maintenance activity and to plan for a certain expenditure on this to be undertaken by routine maintenance gangs working on the road. As noted above, dust control is not a very cost-effective treatment, and social and political judgement will play an important part in determining where and when such activities should be undertaken.

Grading

The relatively rapid deterioration cycles of gravel roads mean that it is not really practicable to undertake conventional condition assessments of the road, and to use these as the basis of undertaking grading as a *condition-responsive* maintenance activity. Grading is therefore normally undertaken several times a year on a *scheduled* basis. The frequency should be chosen to keep the road in as near optimum economic condition as possible.

Optimum economic condition can be conceptualised as shown in Figure 6.8. This shows that as the maintenance standard, or grading frequency, increases, so does the maintenance cost. But improved maintenance standards result in lower road user costs because smoother road surfaces are provided. The sum of the maintenance cost and road user cost curves is the total transport cost, and the resultant curve is 'U'-shaped. The optimum maintenance standard is that which minimises the total

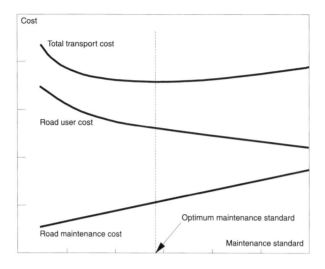

Figure 6.8 *Concept of an optimum maintenance standard*

transport cost. It is this approach that should be used to determine the optimum grading frequency.

An appropriate frequency of grading should be determined for each individual road. Initial frequencies of grading should be carried out on the basis of economic studies using road investment models, such as HDM-III developed by the World Bank (Watanatada *et al.*, 1987). The appropriate level of maintenance should be determined in terms of the life cycle cost of the road; the investment model can be used to search for that frequency of maintenance that minimises the sum of maintenance cost and road user cost over the road's life. The optimum value of maintenance derived in this way will be different for roads in different areas, for roads carrying different levels of traffic, and for roads built with different materials.

Figure 6.9 shows, as an example, a plot of life cycle costs against grading frequency and traffic level derived in this way for particular roads in Papua New Guinea. For the higher traffic levels, distinct minima of total cost can clearly be seen, indicating the optimal grading frequency. Figure 6.10 shows optimal frequencies plotted for a number of different countries, and shows that these optima are quite different for each of the countries where studies have been reported. This emphasises the need to carry out detailed investigations, not only for each country and traffic level, but also for each type of surfacing material.

This approach to selecting grading frequencies has the advantage of producing economic justifications for particular courses of action. However, to employ this method requires good knowledge about the gravel materials being used and the rates at which they will deteriorate under different traffic levels and climatic situations. Where such knowledge is not available, a less sophisticated approach has been proposed (Jones and Robinson, 1986), which has used the data reported in Figure 6.10 to produce the grading frequency chart shown in Figure 6.11. This suggests the range of

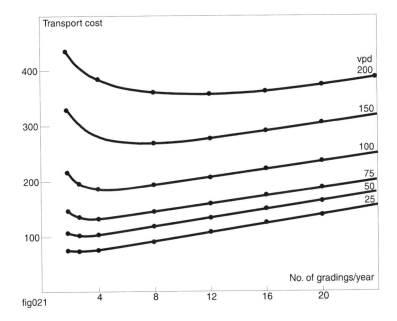

Figure 6.9 *Dependency of transport cost on grading frequency* (*Source*: Jones and Robinson, 1986. Crown copyright 1986. Reproduced by permission of the Controller of H. M. Stationery Office)

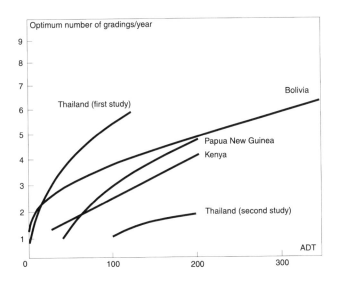

Figure 6.10 *Optimum grading frequency curves for several countries* (*Source*: Jones and Robinson, 1986. Crown copyright 1986. Reproduced by permission of the Controller of H. M. Stationery Office)

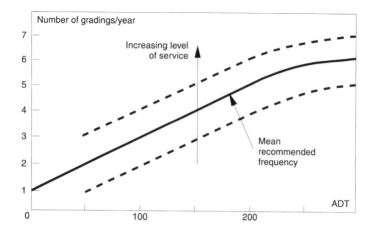

Figure 6.11 *Grading frequency chart* (*Source*: Jones and Robinson, 1986. Crown copyright 1986. Reproduced by permission of the Controller of H. M. Stationery Office)

grading frequencies that are likely to be appropriate for different traffic levels, and indicates how the 'level of service' can be changed by altering the grading frequency. The error bands reflect the relative insensitivity of the total transport cost to variations in grading frequency. As a result, grading frequencies can often be reduced from theoretically optimal values with a significant saving in the road administration's costs, but with only a low probability of an increase in total transport cost.

6.9 Contract packaging

When works are to be undertaken by contract, there may be a minimum size of works for which the use of a contract is cost-effective. Also, where treatment lengths are relatively short, it may sometimes be more cost-effective to combine treatment lengths, even of a different treatment type, into contract packages. A method for contract packaging has been described by Snaith *et al.* (1994).

In this application, decision rules are used to combine defect lengths into treatment lengths on which it is economic for works to be carried out. On any length of road, the aim is to maximise the economic return by grouping projects or schemes of the same or compatible works on contiguous road lengths. This may involve applying more expensive treatments than would otherwise be justified on some sections of road. The process may be considered in terms of seeking economies of scale in terms of treatment cost. The basic logic is illustrated in Box 6.3. The logic is applied iteratively within a framework of rules, and covers all sections and treatments to derive an optimal selection. The same principle can also be applied using NPV instead of treatment cost to enable user benefits to be included.

> **Box 6.3**   *Example of approach to contract packaging*
>
> Consider the three sections of road a, b and c shown below.
>
>
>
> *Contiguous road sections requiring different treatments*
>
> Sections a and c require an overlay at a fixed cost of $\$f_1$, and a variable cost of $\$v_1$ per unit area. Section b requires surface dressing at a fixed cost of $\$f_2$, and a variable cost of $\$v_2$ per unit area. Therefore the cost of applying separate treatments to the three sections is
>
> $$(\$f_1 + \$v_1 \cdot \text{area}_a) + (\$f_2 + \$v_2 \cdot \text{area}_b) + (\$f_1 + \$v_1 \cdot \text{area}_c)$$
>
> However, if an overlay were to be applied throughout, then the cost would be
>
> $$\$f_1 + \$v_1(\text{area}_a + \text{area}_b + \text{area}_c)$$
>
> It follows that if
>
> $$(\$f_1 + \$f_2)/(\$v_1 - \$v_2) \geq \text{area}_b$$
>
> then it is cost-effective to overlay all sections, rather than to overlay a and c, and to surface dress b.
>
> *Adapted from:*   Snaith *et al.* (1994).

7 Prioritisation

- Basis concepts
  - generations of decision-support systems
  - commercial and user models
- First generation prioritisation methods
  - defectiveness indices
  - degree of defectiveness
  - treatment-based methods
- Second generation methods
  - cost-effectiveness methods
- Third generation methods
  - objective function
  - total enumeration
  - economic boundary
  - consideration of uncertainty
- Reduced time period analysis methods
  - theoretical concerns over life cycle analysis
  - the budget period method
  - annualised cost method
- Prediction of deterioration
  - probabilistic methods
  - mechanistic models
  - regression models
  - mechanistic–empirical models
  - calibration

7.1 Basic concepts

7.1.1 Generations of decision-support system

As decision-support systems have evolved over time, they have become more sophisticated in terms of the way that they:

- Subdivide the road network into sections
- Use intervention levels to determine treatments
- Enable the analysis of different numbers of treatment options per section
- Carry out economic analysis
- Prioritise decisions when there are budget constraints.

It is common to group systems as 'first', 'second', or 'third' generation, depending on their approach to addressing the above issues. The general characteristics of each group are described in Table 7.1.

Many proprietary management systems combine features from different system generations. Nevertheless, grouping by system generation provides a convenient framework for describing the different prioritisation methods. Condition projection methods are required by both second and third generation methods, and these are also described.

Table 7.1 *Grouping of management systems by generation*

	First generation	*Second generation*	*Third generation*
Sectioning	Constant length sections, or fixed length sections based on data collection intervals	Pre-defined, variable treatment lengths based on physical pavement characteristics and traffic	Variable treatment lengths obtained by combining defect lengths after the analysis for efficiency of undertaking works ('dynamic sectioning')
Intervention levels	Intervention levels based on present pavement condition plus traffic	Intervention levels consider prediction of pavement condition	Intervention levels based on life cycle prediction of both deterioration and impact on road users
Treatment options	One standard treatment prescription per section	Comparison of do something and do minimum treatment options for each section	Consideration of multiple treatment options per section
Basis of economic analysis	Present cost of treatments	Present and future costs of treatments and benefits to road administration ('commercial' models)	Life cycle approach to costing and inclusion of benefits to road users ('user' models)
Method of prioritisation	Ranking based on function of present costs, condition and road hierarchy	Ranking based on cost-effectiveness, with consideration of treatment life, and analysis of deferment options	Formal optimisation of multiple treatment options per section over a multi-year period

Setting priorities for treatments to gravel roads is, in principle, no different from that for paved roads. However, where scheduled treatments are used, such as has been recommended for grading, these will need to be dealt with outside the formal prioritisation process; frequencies of grading will need to be scheduled to match the budget level that is available for this particular activity. It may be necessary to preassign a budget for this. Other than this, where works on gravel roads are to be carried out under the same budget head as that for paved roads, they can be considered within the same prioritisation process. Any of the methods described under first, second or third generation can be used.

7.1.2 *Commercial and user models*

Several of the prioritisation methods are based on considerations of cost. For example, priorities may be selected in such a way that minimises present and future cost.

Road works costs can be considered under two headings:

- *Road administration costs*
  - Costs of the actual work (equipment, labour, materials, overheads, etc.), which are often on an in-place basis (volume, area, mass), such as cost per lane-kilometre
- *Road user costs*
  - extra vehicle operating costs during the works
  - time costs for road users and, in particular the cost of user delays during works
  - accident costs due to traffic hazards or interruptions associated with the works
  - environmental damage (air or water pollution, noise, etc.).

Future costs need to be discounted to give their present worth if the values of different options are to be compared. Two different types of economic model can be used for the analysis: 'commercial models' and 'user models' (Thagesen, 1996).

Commercial models include only costs to the road administration. Future works, as well as present works, are included in the costings. Future works costs required to meet target standards are either estimated subjectively, or calculated objectively using predictions of future conditions of the roads. Future costs are discounted back to the present year.

User models include costs to road users as well as those to the road administration. Although user costs are not part of an administration's budget, it is the user who ultimately pays the cost of maintenance and delay, either directly or indirectly. Some systems, therefore, consider it appropriate for these costs to be included in the prioritisation process.

7.2 First generation prioritisation methods

7.2.1 Basic principles of the methods

First generation decision-support systems are characterised by prioritisation methods where ranking is based upon present values of condition, included in an *ad hoc* function along with costs and road class. Normally, one treatment per section is identified, and all sections where treatments are needed are listed in priority order to determine where the budget cut-off line should be drawn. Treatments that cannot be funded when budgets are constrained are deferred at least until the next year, when they will be reconsidered along with other treatments at that time.

First generation methods can, in principle, be used for planning, programming and preparation, and at both network and project level. However, most methods described were derived typically for programming applications to select projects to be undertaken within the anticipated budget.

It is convenient to discuss first generation methods under three headings:

- Defectiveness index
- Degree of defectiveness
- Treatment-based methods.

7.2.2 Methods based on defectiveness indices

Early approaches to road network management used indices to characterise the condition of sections of road. The use of 'defect indices' provides a mechanism that enables defects of different types to be combined to determine a 'priority index'. These indices are sometimes known as 'serviceability indices', and are used both for treatment selection and for prioritisation. Even now, many first generation systems still make use of such indices, although other parameters may also enter the decision-making process (American Society of Civil Engineers, 1992).

The most common index is the *Present Serviceability Index* (PSI), but there are several others that are similar in concept (Haas *et al.*,1994). PSI ranks condition on a scale of 1 to 5, where 5 represents excellent condition. Its value is determined by engineers using visual assessment, although a relationship between PSI and condition measurements was established at the time of the AASHO Road Test (Liddle, 1963), in terms of 'slope variance' (roughness), rutting, cracking and patching.

An example of defectiveness index methods is the *Rational Factorial Rating Method* (Haas *et al.*, 1994), where expert opinion was used to develop a priority index. The method is illustrated in Box 7.1.

Box 7.1   *Rational Factorial Rating Method*

Priorities assigned by a panel of engineers were analysed statistically to develop the following equation for prioritisation:

$$Y = 5.4 - 0.0263(X1) - 0.0132(X2) - 0.4\log(X3) + 0.749(X4) + 1.66(X5)$$

where
    Y = priority index in which a value of 1 represents very poor condition and a value of 10 represents excellent condition (thus a low value indicates that a pavement should have a high priority for treatment)
    $X1$ = rainfall (5 to 40 inches per year)
    $X2$ = freeze and thaw (0–60 cycles per year)
    $X3$ = traffic (100–100 000 ADT)
    $X4$ = PSI (4.0–2.0)
    $X5$ = distress rating (+1.0 to −1.0)

Source: Haas *et al.* (1994).

7.2.3 Methods based on degree of defectiveness

In this group of methods, priorities depend upon the amount by which a measured condition or defect rating exceeds an intervention level: the more the intervention level is exceeded, then the higher the priority. Typically, 'condition indices' are determined for individual defects, and these are combined in a functional relationship to give a 'priority index'. Two basic methods exist for determining defect indices. The first uses the difference between the defect value and the intervention level:

$$CI_i = D_i - A_i$$

where CI_i = value of the condition index for defect i
D_i = condition measurement or rating of defect i
A_i = intervention level indicating the acceptable amount of defectiveness before a treatment is warranted for defect i.

The second method uses the ratio of the defect value to the intervention level:

$$CI_i = D_i/A_i$$

with notation as above. Priorities are normally derived using a relationship of the form:

$$P = \sum_i k_i CI_i$$

where P = priority index
k_i = constant
and other notation as above.

Examples of the approach using the 'difference method' are the CHART and MARCH systems used in the United Kingdom (Atkinson, 1997).

In the 'ratio' method, priority indices are often expressed as a percentage:

$$P = 100 \sum_i k_i CI_i$$

In this case, a priority index of 100 per cent indicates that a length of road has the same defectiveness as the intervention level; thus, the treatment is just warranted. Similarly, an index of 145 per cent indicates that there is 45 per cent more defectiveness than in the intervention level. Defect measurements or ratings may refer to an individual defect, such as *whole carriageway major cracking*, or to a 'collective' defect, such as *minor carriageway surface deterioration*, which is a combination of cracking, crazing, ravelling and the like. In all cases with this method, rules are needed to combine the condition indices of individual defects into priority indices. An example of the use of rules is given in Box 7.2.

With both of these methods, the effects of other parameters, such as road hierarchy, traffic, environment, road construction type, etc., are normally taken into account by setting intervention levels to different values to reflect the differing requirements. Some systems, such as 'BSM', also formalise the setting of intervention levels by basing them on life cycle costing considerations through linkages to HDM-III, as shown in Figure 7.1.

Both the 'defectiveness index' and the 'degree of defectiveness' methods usually base priorities purely on the basis of the severity of condition; generally no account is taken of the treatment that will be applied or the traffic, other than through the value of the intervention level. Such methods are sometimes referred to as the 'worst first' approach.

Box 7.2 *Example of an application of the 'degree of defectiveness' methods*

The particular method works in the following manner. Three severity levels are defined for each defect rating: low, medium and high. A condition index is determined for each severity level, as follows:

$$CI_i = 100 + [(D_i - A_i) \times G_i]/A_i$$

where D_i = rating of defect i
A_i = target or acceptable rating of defect i
G_i = weighting factor.

For each defect, the condition indices for the three (low, medium and high) severity levels are added together to give a combined condition index. The defect giving the highest combined condition index is then selected as being the principle defect, and secondary and tertiary indices are also identified.

The principle, secondary and tertiary condition indices are combined and weighted to provide a priority index for a sub-section:

$$PI = (CI_1 + f_2 CI_2 + f_3 CI_3) \cdot w$$

where CI_j = condition index for principle ($j=1$), secondary ($j=2$), and tertiary ($j=3$) defects
f_j = combination factors established by customisation
w = weighting (situation) factor obtained from a matrix for C, T and E
where C = road hierarchical class
T = traffic volume
E = environment.

Adapted from: RoadFax.

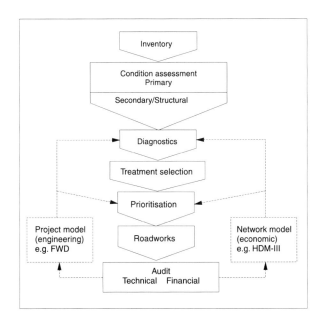

Figure 7.1 *Example of road management system linking to external economic model*

7.2.4 Treatment-based methods

Treatment-based methods determine priorities on the choice of treatment to be applied, rather than simply on the condition of the road. These methods assume that there is a hierarchy among maintenance activities in terms of their impact on road preservation, and this is combined in some way with a traffic hierarchy to provide the basis for setting priorities. The traffic hierarchy reflects the significance of the particular road as a transport link which, in effect, is a surrogate for the road user benefits resulting from a maintenance treatment.

An example of such a method is described in Box 7.3. This method of prioritisation is an example of a pragmatic attempt to reflect the likely rate of return from undertaking different activities, without being based upon strict economic criteria. As such, the method is designed to ensure that every road in the network receives at least the minimum maintenance needed to keep it operational while, at the same time, focusing reactive, preventive and resurfacing works on the economically important roads with high traffic levels. Note that, in this example, strategic roads have absolute priority for resources, including those for capital works.

Box 7.3 *Example of a treatment-based method*

In this method, priority is determined using a matrix linking hierarchies of maintenance activities and road class. The hierarchy of maintenance activities is shown below.

> 1 Emergency work
>   - emergency repairs to blocked or impassable roads
>   - removal of debris and the stabilisation of side slopes
> 2 Cyclic drainage work
>   - cleaning out and re-cutting ditches and turnouts
>   - cleaning out bridges and culverts
>   - filling scoured areas
>   - building check-dams and scour controls
>   - repair of drainage structures
> 3 Reactive work on pavement
>   - patching
>   - local sealing
> 4 Periodic preventive and resurfacing work
>   - resealing
> 5 Other cyclic and reactive work
>   - filling on shoulders and slopes
>   - grass cutting
>   - cleaning, repainting, repairing and replacing road furniture
> 6 Periodic overlay and pavement reconstruction
>   - overlaying
>   - pavement reconstruction

The method considers that emergency works demand top priority to prevent roads being cut. Similarly, cyclic drainage work deserves high priority because neglected drainage can lead rapidly to deterioration of the whole road; repairing surface defects caused by poor drainage is a waste of time and effort unless the drainage is put right first. Periodic work is viewed as a series of discrete projects that have to compete for the resources available and which can be undertaken separately, deferred or brought forward, as required. In this particular example, overlay and pavement reconstruction works are treated as capital projects whose funding does not come from the maintenance budget: this is to avoid major overlay and reconstruction schemes swallowing up the majority of the maintenance budget, leaving little for cyclic, reactive and preventive works.

The method also classifies roads according to their importance: those carrying the heaviest traffic loads normally being the most important in the network from an economic standpoint. These are also the roads most liable to deteriorate rapidly if defects remain untreated. There may also be roads with relatively low levels of traffic which, nevertheless, have key strategic importance because of the places that they link. The method assigns top priority for maintenance work to these, since it is considered vital to keep strategic roads in good condition. The remainder of the network is classified by level of traffic on each road, and the traffic hierarchy recommended by the method is shown below.

Road hierarchy	Traffic range (vehicles/day)	Surface type
1	(Strategic roads)	Paved
2	Greater than 1000	Paved
3	500–1000	Paved
4	200–500	Paved
5	Greater than 200	Unpaved
6	Less than 200	Paved
7	50–200	Unpaved
8	Less than 50	Unpaved

Priorities are determined using the matrix shown below. Maintenance activities are numbered from 1 (highest priority: emergency maintenance on strategic roads) to 48 (lowest priority: capital works on roads with low levels of traffic). Although a structure for the matrix is recommended in the example, the aim is that users will define their own priorities to reflect local circumstances within this general framework.

Hierarchy of maintenance activity	Priority							
	Traffic hierarchy							
	1	2	3	4	5	6	7	8
Emergency	1	7	8	9	10	11	12	13
Cyclic drainage	2	14	15	16	17	18	19	20
Reactive pavement work	3	21	24	27	30	33	36	39
Periodic preventive	4	22	25	28	31	34	37	40
Other cycl/react	5	23	26	29	32	33	34	35
Overlay/ reconstr	6	42	43	44	45	46	47	48

Adapted from: TRRL Overseas Unit (1987). Crown copyright 1987.
Reproduced by permission of the Controller of H. M. Stationery Office.

7.3 Second generation methods

7.3.1 Basic principles of the methods

A drawback with threshold models of prioritisation is that they do not take into account the future consequences of decisions that are taken. The key feature of second generation methods is that they overcome this problem by basing priorities on considerations, not only of present costs, but also on the implications of decisions in terms of future road conditions and costs.

Decisions are based on an assessment of the predicted consequences of carrying out, or not carrying out, treatments identified as being necessary by applying defined intervention levels. Future consequences are considered in terms of costs and benefits of each alternative. One alternative is to 'do nothing' or, as noted earlier, to 'do minimum'; the other is to 'do something'. The needs and consequences of undertaking works on the network are then determined for each road section, and costs and benefits compared for each alternative to determine which has greater priority.

Second generation methods have often been developed specifically to deal with planning and programming activities at network level, where economic decisions are being made about optimal budget levels and network conditions. However, the methods are also appropriate at preparation and project level.

Second generation prioritisation methods consider both present and future costs, and use the criterion of cost-effectiveness, or cost-performance surrogates, to determine benefits. This differs from the full life cycle costing methods which characterise third generation systems. However, the difference between second and third generation methods can be somewhat blurred.

7.3.2 Cost-effectiveness methods

Most second generation prioritisation methods use the concept of cost-effectiveness. In this context, 'effectiveness' is the measure of the future 'worth' of the works that are undertaken. 'Cost' is the present-day cost of the works. Cost-effectiveness is, simply, the ratio of effectiveness to cost, and can be considered as a surrogate for the NPV/cost ratio (see Chapter 4). The ratio has no physical or economic meaning *per se*, but it can be used in the relative comparison of options.

Simple methods

In the simplest cost-effectiveness methods, 'effectiveness' is defined as the expected life of the treatment alternative in years. Cost is determined from the following:

$$\text{Unit cost} = [\text{Cost of (equipment + labour + materials)}] / (\text{Accomplishment or production})$$

Thus:

    Cost-effectiveness = (Expected life in years of the treatment alternative) / (Unit cost)

Often, in such an analysis, the value quoted is the inverse of cost-effectiveness, which is the 'average annual cost'. Thus, the treatment alternative with the lowest average annual cost represents the most cost-effective solution. In this simple example, future consequences of actions are only taken into account through the inclusion of the expected life of the treatment. There is no attempt to discount costs to reflect their time value. As such, methods of this type should really be classed as 'first generation'. However, the example serves to illustrate the general principle of the cost-effectiveness type of methods.

Simple cost-effectiveness methods tend to give priority to the cheaper treatments, such as surface dressing, because these give good returns in terms of increased life per unit of cost, but does not take account of consequential costs. However, because of the impact of mobilisation costs, the size of works also affects the cost-effectiveness (see discussion of contract packaging in section 6.9).

Performance curves

In general, second generation cost-effectiveness methods make use of a conceptual 'performance curve' to determine the effectiveness of a particular treatment. These curves recognise that an aim of the road manager is to provide the best possible pavement condition for as long as possible. Effectiveness can therefore be defined as the net area under the pavement condition–time curve multiplied by the volume of traffic for a road section of unit length. This concept is illustrated in Figure 7.2, and effectiveness (E) is given by the following:

$$E = (ADT) \, t \sum_{i=1}^{t} P_i$$

where P_i = road condition at time i
       ADT = average daily traffic
       t = treatment life, normally in years.

The method of determining effectiveness is illustrated further in Figure 7.3. This compares the effectiveness of treatments with different lives t_a and t_b when applied on the same section of road. The difference in the areas under the curves gives the increased effectiveness of treatment b over treatment a. A further example, illustrated in Figure 7.4, compares the effectiveness of treatments on different sections. The increased effectiveness of treatment b over treatment a is given by the *difference* in the shaded areas under the graph, as shown in the figure. This approach enables choices to be made of a package of treatments to implement across a network when budgets are constrained. In both cases, treatment choices would be determined by dividing the difference in effectiveness by the difference in cost to calculate the cost-effectiveness.

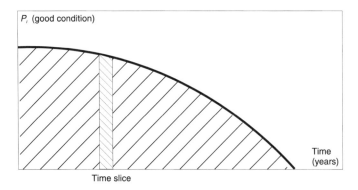

Figure 7.2 *Conceptual pavement performance curve*

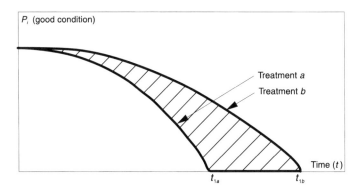

Figure 7.3 *Performance curve as a result of applying alternative treatments to a road section*

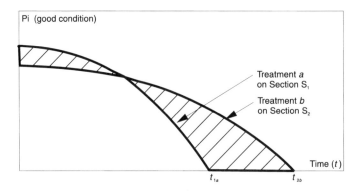

Figure 7.4 *Performance curve as a result of applying alternative treatments on different sections*

An example of a cost-effectiveness approach using a performance curve is described in Box 7.4.

Box 7.4   *Example of a second generation cost-effectiveness approach using a performance curve*

In this example, a treatment is delayed until later than the year in which its need was first identified. The effect of this treatment intervention is shown, as in the continuing deterioration thereafter in the figure below. Effectiveness is given by the net area in the shaded area to the right of the works implementation year minus the shaded area to the left.

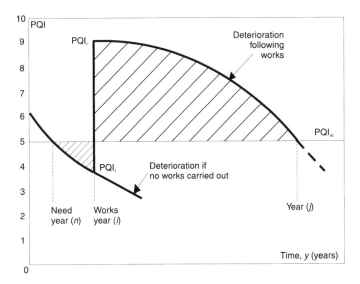

In this example, effectiveness is given by:

$$E = Area \times ADT \times length$$

where

$$Area = \sum_{y=i}^{j}\left(PQI_y - PQI_m\right)y - \sum_{y=n}^{i}\left(PQI_m - PQI_y\right)y$$

and PQI_y = Pavement Quality Index in year y
  n = year in which treatment is first needed
  i = year in which treatment is implemented
  j = year when pavement condition returns to the minimum acceptable level following treatment.

Adapted from: Haas *et al.* (1994).

7.4 Third generation methods

7.4.1 Statement of the problem

The essential components of third generation systems that distinguish them from others are the inclusion of formal optimisation of multiple options, and full life cycle analysis of both road administration and user costs. These features differentiate third generation systems from the cost-effectiveness methods, described in the previous section, although the techniques used by proprietary systems are often similar. The prioritisation problem can be considered as aiming to choose that set of treatment options on sections in the network, over time, that meet certain pre-defined 'prioritisation criteria'. On any network, there will be:

n sections
m treatment options per section, and
t years of analysis.

A typical solution to this problem is shown in Table 7.2.

Table 7.2 *Typical result from a third generation prioritisation method*

Section	Year				
	1	2	3	4	t
1	a_{10}	a_{10}	a_{12}	a_{10}	a_{10}
2	a_{22}	a_{20}	a_{20}	a_{20}	a_{23}
3	a_{31}	a_{30}	a_{30}	a_{32}	a_{30}
4	a_{40}	a_{41}	a_{40}	a_{40}	a_{42}
n	a_{n0}	a_{n0}	a_{n3}	a_{n0}	a_{n0}

Where, for example:
a_{i0} = the do minimum treatment of routine maintenance only
a_{i1} = patching
a_{i2} = surface dressing
a_{i3} = 50 mm overlay
a_{i4} = 150 mm overlay
a_{i5} = pavement reconstruction.

7.4.2 Objective function

The prioritisation criteria are encapsulated in mathematical terms by an 'objective function' that is maximised or minimised by the method. The process of maximising or minimising an objective function is known as 'optimisation'. Typical aims of the prioritisation that need to be defined by an objective function are to find the solution, in terms of the treatment options for each section for each year of the analysis, that:

- Maximises the NPV subject to a given budget constraint, where NPV is typically defined as
    the discounted life cycle costs to the road administration and to the road users of doing only routine maintenance on all sections in the network (do minimum)
    minus
    the discounted life cycle costs to the road administration and to the road users of undertaking routine maintenance and a set of treatment options on sections in the network (do something).
- Minimises the sum of costs to the road administration and the road user cost over time; this finds the optimal budget for each year (unconstrained optimisation).
- Minimises the road user costs over time subject to constraints of
    – minimum road standard
    – maximum annual budget
- Maximises road conditions over time subject to a given budget constraint.

Maximisation of NPV under budget constraint is the objective function that is most widely used, and this will be taken as the basis of further discussion.

The prioritisation method is, thus, aiming to choose between treatment options by optimising an objective function.

7.4.3 Total enumeration

One way of solving the prioritisation problem is simply to evaluate all possible solutions and then to choose that which optimises the objective function. This is known as 'total enumeration', and the approach is defined in mathematical terms in Box 7.5 where the *Expenditure budgeting module* of the World Bank's HDM-III model (Watanatada *et al.*, 1987) is described. In general, there are $m^{n.t}$ (using the above notation) solutions that need to be evaluated by such a method. Thus, for a road network of ten sections ($n = 10$), each with six treatment options ($m = 6$), analysis for one year only ($t = 1$) requires 6^{10} evaluations of the objective function, which represents over 60 million calculations. Using this method to produce a ten year plan for even a relatively small network of, say 5000 sections, would require $6^{50\,000}$ evaluations. Such computational requirements are impracticable. Total enumeration is, therefore, only applicable where the size of the prioritisation problem is relatively small.

7.4.4 Dynamic programming

Two other solution methods can be used to overcome this computational problem. The first is known as 'dynamic programming' (Bellman and Dreyfus, 1962). This uses the basic assumption that any optimal solution can consist only of optimal sub-solutions. Thus dynamic programming divides the optimisation problem into a number of smaller problems that are easier to solve. Instead of examining all possible solutions, dynamic programming examines a small carefully chosen sub-set, rejecting those combinations that cannot possibly lead to an optimal solution. An advantage of dynamic programming is that, once a problem is solved for a long analysis period, results are readily available for shorter periods (Shahin, 1994). The application of this approach to prioritisation is described by Watanatada *et al.* (1987). This method is not discussed in any more detail here.

Box 7.5 *The HDM-III Expenditure Budgeting Model approach to prioritisation using total enumeration*

The optimisation problem is defined as one of maximising the total net present value for all treatments and all links in the network during a particular budget period:

$$\text{Maximise } TNPV[X_{km}] = \sum_{k=1}^{K} \sum_{m=1}^{M_k} NPV_{km} X_{km}$$

where k = a road link
   m = a treatment option for a link
   K = the number of links in the road network
   M_k = the total number of treatment options
   NPV_{km} = the net present value of undertaking treatment option m on link k relative to a 'do minimum' case

subject to resource constraints:

$$\sum_{k=1}^{K} \sum_{m=1}^{M_k} R_{kmqt} X_{km} \leq TR_{qt}, \quad q = 1, \ldots, Q; t = 1, \ldots, T$$

where R_{kmqt} = the (undiscounted) budget under budget head q incurred by the road administration in a budget period t, where
   Q = the total number of budget heads
   T = the total number of budget periods (the duration of t may be one or more years and need not be equal for different budget periods)
   TR_{qt} = the maximum budget available under budget head q in budget period t

and, to ensure that no more than one treatment option m can be implemented on any one link k, the following constraints of 'mutual exclusivity' are also needed:

$$\sum_{m=1}^{M_k} X_{km} \leq 1, \quad k = 1, \ldots, K$$

Adapted from: Watanatada *et al.* (1987). © The International Bank for Reconstruction and Development /The World Bank.

7.4.5 Economic boundary

Definitions and assumptions

The second approach to solving the problem is an 'heuristic' method. This class of method makes use of 'unconventional' means to arrive at the correct solution. The method is described variously as the 'effective gradient' method (Ahmed, 1983), and the 'incremental NPV/cost' method (Phillips, 1994), as well as the 'economic boundary' or 'efficiency frontier' method. It is described below in more detail.

It is necessary to distinguish between types of 'option': 'treatment options' are the range of possible maintenance treatments that can be applied to any section; a 'solution option' is one set of treatment options, on each section, for each year of the analysis, for which an objective function can be evaluated. The set of treatment options shown in Table 7.2 is, therefore, one solution option. Assume that the aim is to maximise NPV for the network subject to a budget constraint. The value of the objective function will then be the NPV of the particular solution option relative to the 'do minimum' solution option of routine maintenance only on all sections in all years.

Optimisation of options on one section only

Consider the solution options on one section only considered for just one year of analysis. In this particular case, a treatment option is identical to a solution option. An example of this case is illustrated in Figure 7.5, where the NPV of each option is plotted against its cost. Note that the options are mutually exclusive, as defined in Chapter 4. The optimisation aims to choose between options on the basis of economic efficiency. The most efficient solution is that which gives the highest NPV per unit of cost expended. Figure 7.6 shows that Option B is the most efficient solution because it has the maximum NPV/cost, represented by the steepest gradient from the do minimum solution. Provided there is sufficient budget to fund this option, then it would be selected as a potential solution to the prioritisation problem. If there was insufficient budget to fund this option, then Option A could be selected instead as a cheaper option.

The method then considers whether it is possible to improve on the selected option by finding a solution option with a higher NPV. It is now, therefore, necessary to consider incremental improvements in NPV that might be possible for incremental increases in cost. However, first it should be noted that certain of the options are non-

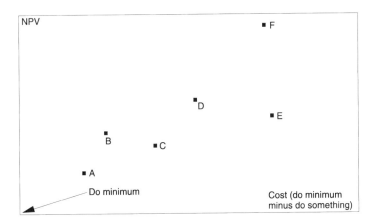

Figure 7.5 *Example of NPVs and treatment costs on one section*

viable. For example, Option C is non-viable because, despite having a higher cost than Option B, it has a lower NPV. This is illustrated in Figure 7.7. The incremental NPV divided by the incremental cost of Option C relative to Option B is negative. Option C would never be selected as a potential solution option under any circumstances. The same applies to Option E. These non-viable options can be removed from further consideration at an early stage in the analysis.

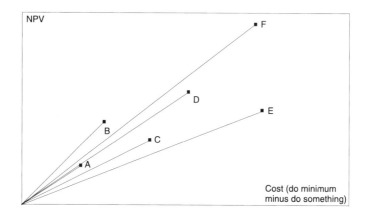

Figure 7.6 *Efficiency characterised by gradient*

The best incremental solution option relative to Option B, in this case, is Option F since the incremental NPV/cost is greatest. In other words, it has the steepest gradient relative to Option B, as shown in Figure 7.8. Option F would be selected instead of Option B if there were sufficient budget. If not, Option D would be considered.

The envelope of all viable options is known as the 'economic boundary'. This is shown for this example in Figure 7.9. Essentially, the prioritisation method searches along this to find that solution option with the highest NPV that can be funded from the available budget. Note that when the economic boundary is concave, as for some options in Figure 7.9, then finding a solution becomes difficult. Some methods in use do not have the facility to do this.

Network analysis

The example described was for several treatment options, but for only one section and only one year of analysis. Extending the prioritisation problem to a multi-year analysis of a network of sections follows exactly the same principles. However, in the more general case, 'treatment options' are replaced by 'solution options' in the analysis. Thus, each point on the NPV versus cost graph represents a solution option, such as the set of treatments shown in Table 7.2, and the economic boundary can be represented by a 'surface' with the third dimension being the year of analysis.

The algorithmic process used to solve the problem is relatively complex, and has been presented here in a simplified graphical form to aid understanding. For example,

Prioritisation 205

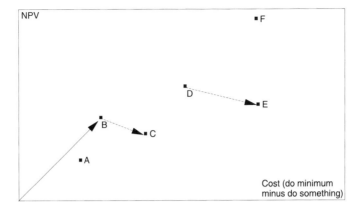

Figure 7.7 *Non-viable options*

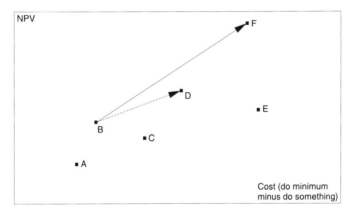

Figure 7.8 *Selection on basis of incremental NPV/cost*

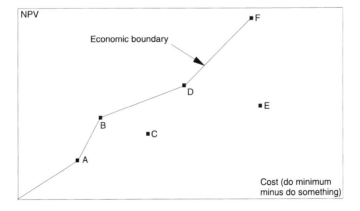

Figure 7.9 *Concept of the economic boundary*

the details of the method that searches along the economic boundary for solution options differs depending on whether the economic boundary is 'convex' or 'concave' at the present search point.

Although this approach to prioritisation is widely used, the analysis of real networks on a multi-year basis still presents a formidable computational problem. For this reason, simplifications are often sought, and some of these are discussed below.

7.4.6 Year-by-year analysis

The economic boundary method of prioritisation is used by several proprietary systems for the programming function. In this context, the aim is normally to select those treatment options over the network that can be funded in the next budgeting period. Because only one year of analysis is being considered, the size of the computational problem is reduced significantly to m^n, with notation as above.

One way of minimising the computational requirements in the multi-year analysis case is to solve the prioritisation problem on a year-by-year basis. In this approach, treatment options are selected across the network for the first year of analysis only. These are then 'fixed', and the second year analysis takes the consequences of the first year treatments as its starting point. Each year's treatments are 'fixed' in turn in this way. This approach is used widely to produce a rolling programme of works, typically over a three year period. The programme of works for the next budget year is, therefore, known with some certainty, but it is recognised that subsequent years' works are known less precisely. The rolling programme is updated every year, so any errors in the longer term predictions are corrected over time. The aim of this approach is to provide only notional works programmes for later years.

The approach can be extended to multi-year planning. It must be recognised that the result produced is sub-optimal. However, since the error is greatest in the later years of the analysis, the impact of any uncertainty on current decisions is reduced. The approach has significant impact on the size of the computational problem, reducing it from $m^{n.t}$ to $t.m^n$ potential solutions. Nevertheless, the size of the multi-year planning problem still means that it needs to be undertaken using only representative sections and works 'categories', rather than works 'types' or 'activities', as noted in Chapter 1.

7.4.7 Consideration of uncertainty

Clearly, prioritisation of multiple treatments, sections and years involves considerable uncertainty in the result. Given the size of the computational problem, it is unlikely to be feasible to carry out any kind of risk or sensitivity analysis on the findings. Uncertainty is normally dealt with by re-running the planning or programming analysis on an annual basis, so at least the forecasts for the next budget period have a lower uncertainty, as discussed in the previous sub-section.

A novel way of dealing with some aspects of uncertainty have been incorporated into the dTIMS prioritisation software produced by Deighton Associates. This considers the uncertainty in costs and NPVs of those solution options lying close to the economic

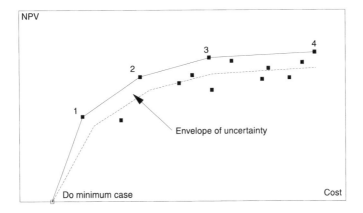

Figure 7.10 *Uncertainty envelope around the economic boundary*

boundary. Recognising the existence of uncertainty means that some treatment options lying close to the economic boundary should also be included in the analyses. To cope with this, an 'envelope' can be created around the bottom of the economic boundary, and treatment options falling within this are included in the analysis (see Figure 7.10). This increases the number of options available for selection, and can also result in producing a 'smoother' annual budget requirement than the deterministic approach.

A second uncertainty consideration concerns incorporation of the marginal treatment option at the budget cut-off. For some options, there can be only a marginal increase in the NPV for a significant increase in the treatment cost. In these cases, the incremental *NPV/cost* ratio approaches zero. Given that there is much more uncertainty in the estimates of NPV, which are based on road user cost, than in the works costs, there is a possibility of mis-allocation of resources. A cut-off can be incorporated into the analysis to avoid this potential problem. When the incremental *NPV/cost* ratio falls below this cut-off, the works option is excluded from the analysis, even if it falls on the economic boundary.

7.5 Reduced time period analysis methods

7.5.1 *Theoretical concerns over life cycle analysis*

In the case of programming, where the concern is to determine the works to be carried out in the next budget year, it has been argued that the use of a full life cycle approach to prioritisation is actually flawed (Robinson and Phillips, 1992). In this case, the basic issue is to decide whether to carry out a treatment this year (do something), or to defer undertaking works until next year (do minimum). It is argued that considerations of cost, traffic, treatments and the like, which might be undertaken 10 to 20 years into the future, should not affect this decision. Unlike project appraisal, where a once-and-for-all decision is taken, decisions on treatment priorities are taken on an annual basis, and that full life cycle analysis may not be appropriate.

In particular, optimising over long time periods poses several difficulties:

- There is considerable uncertainty surrounding traffic flow projections that must be made for long periods into the future to cover the whole analysis period.
- Traffic growth may well lead to congestion in later years of the analysis, which distorts the decision-making process.
- The identification of future maintenance treatments and user costs relies on projection of pavement condition in both the 'do something' and 'do minimum' cases; these need to be based on the projected traffic, and require that assumptions must be made about the effect and durability of maintenance treatments over the whole analysis period.
- When future maintenance treatments are identified and costed for the calculation of NPV, assumptions must be made that each of these treatments is economically justifiable at the time it is needed.
- Life cycle analysis assumes that sufficient budget will be available to fund all future works, but budget levels for several years ahead are seldom committed, and are subject to considerable uncertainty.

These are all major assumptions and call into question the validity of using such long periods for the analysis of treatment options under constrained budgets. It is for these reasons that several prioritisation methods now use criteria other than NPV or life cycle costs for the objective function. These methods also have the advantage that computational requirements are reduced significantly, and thus provide an additional mechanism for addressing the problems in this area that were highlighted earlier.

7.5.2 Basis of the approaches

There are two main types of reduced time period analysis methods:

- Budget period analysis
- Annualised cost method.

The first is essentially the same as the life cycle analysis method, but uses a shorter analysis period, typically of five years, known as the 'budget period'. The second method considers costs and benefits in the first year only, but 'annualises' these over the life of the treatment being considered to reflect the impact on the benefits of using treatments with different lives.

7.5.3 The budget period method

This method is used by the United States Federal Highway Administration in their *Highway Economics Requirements System* (HERS), and is typical of this class of method. For example, a similar approach has been incorporated into HDM-4. Discounted costs and benefits are evaluated only for a short analysis period, known as a 'budget period' or 'funding period'. This is considered to be more realistic, particularly when available funding is only known for a few years. There are two particular issues to be considered with this approach:

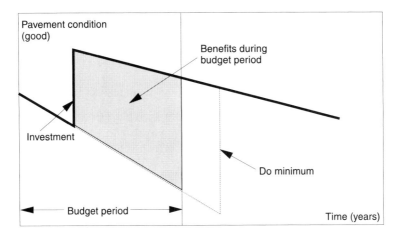

Figure 7.11 *Determination of benefits in the budget period analysis method*

- The determination of the residual value of the investment at the end of the short budget period.
- The estimation of benefits accruing due to the value of the investment beyond the budget period.

One approach to taking account of these issues is to estimate benefits by comparing treatment costs incurred during the budget period with a corresponding 'do minimum' case, where the same treatment is deferred to the year immediately after the budget period. This is illustrated in Figure 7.11. An example of this would be an overlay applied in the second year of the budget period, compared with a similar overlay carried out in the year after the budget period. A simplifying assumption is normally made that works carried out after the budget period will be less effective in improving pavement condition. As a result, a more significant treatment will be needed to restore the condition to the same as that which would have resulted from the 'do something' works, as shown in Figure 7.11. It is also assumed that pavement performance after the budget period will be similar for both alternatives. With these assumptions, the residual value issue is resolved since there is no need to calculate pavement deterioration and road user costs beyond the first year after the budget period. This also reduces computational requirements significantly.

7.5.4 Annualised cost method

In this method, instead of maximising the NPV, an objective function is used that does not require the calculation of life cycle costs. 'Total annualised system costs' (TASC) are used instead (Phillips, 1994). These consist of the sum of the costs to the road administration and the road user as a result of undertaking a particular treatment, annualised over the life of the treatment. Annualising costs over a number of years gives the cost in each year that would be equivalent to a single cost at the start of the period. Without discounting, this just requires dividing by the number of years but, with discounting, the annualising function needs to be used. This is:

$a(r, t) = r / [1 - (1 + r/100)^{-t}]$
where r = discount rate (per cent)
 t = number of years over which the cost is annualised.

As r tends to zero, so $a(r, t)$ tends to $1/t$. Thus, for example, using a discount rate of 8 per cent, a $1000 cost each year for 10 years is equivalent to a cost of approximately $7600 now or, conversely, a $1000 cost now is equivalent to a cost of approximately $150 a year for 10 years.

Thus, instead of using NPV in the optimisation, a surrogate is used of the form:

$$\sum (TASC_M - TASC_S)$$

where $TASC_M$ = annualised cost of doing routine maintenance only
 $TASC_S$ = annualised cost of undertaking a particular treatment

with the summation being over all sections of the network. This function can be used in the prioritisation, with significant savings in computational effort.

The method is an attempt to evaluate the consequences of investing in the present works only, without the need to consider works that may or may not be needed at periods in the future. Annualising is an appropriate economic measure to reflect the future discounted values of benefits, and enables the value of the different lives of treatments to be taken into account. The approach also obviates the need to determine residual values. Case studies have shown that the approach gives similar rankings to *NPV/cost*, although there are small differences in the position of individual projects or schemes in the rankings produced. It can, therefore, be used with some confidence.

7.6 Prediction of deterioration

7.6.1 *Approaches to condition projection*

Normal methods of maintenance assessment involve condition surveys being carried out in one year which are used as the basis of maintenance treatments to be applied in the following year. The quality of decision-making with such an approach could clearly be improved by projecting pavement condition forward one year from the time of assessment so that decisions on treatments could be based on road conditions expected to pertain at the time the treatment is carried out, rather than those at the time of assessment. Alternatively, the projection could be applied to the intervention levels. Condition projection methods are also used either explicitly or implicitly by all second and third generation methods of prioritisation, since knowledge is needed of pavement conditions in the future, with and without maintenance treatment.

A number of different condition projection methods exist, and these have been categorised by Haas *et al.* (1994) into four types:

- *Subjective*
  Where experience is 'captured' in a formalised or structured way to develop deterioration prediction models.
- *Purely mechanistic*
  Based on some primary response or behaviour parameter, such as stress, strain, or deflection.
- *Regression*
  Where the dependent variable of observed or measured deterioration is related to one or more independent variables, such as subgrade strength, axle load applications, pavement layer thicknesses and properties, environmental factors and their interactions.
- *Mechanistic–empirical*
  Where the form of the required equation is defined, based on mechanistic principles, to relate a dependent variable to measured deterioration, such as surface distress or roughness; regression analysis is then used to determine the actual coefficients and parameters within this pre-defined form.

These can also be grouped into two basic classes:

- *Probabilistic*
  Where condition is predicted as a probability function of a range of possible conditions.
- *Deterministic*
  Where condition is predicted as a precise value on the basis of mathematical functions of observed or measured deterioration; this class includes
  – mechanistic
  – regression
  – mechanistic–empirical.

Examples of different approaches to performance prediction have been reported by the Transportation Research Board (1994).

7.6.2 Probabilistic methods

This class of model typically predicts the probability of a particular road condition prevailing at a fixed time in the future, the value of which depends only on the current condition. The probability levels are sometimes assigned to the possible future outcomes by engineering judgement, and these are often determined by the analysis of estimates made by a panel of expert engineers. Better methods assign probabilities on the basis of an analysis of past performance of sections in a road network (Kerali and Snaith, 1992; Jackson *et al.*, 1996).

Two types of probability functions can be used (Shahin, 1994). The first is a probability distribution, which is a continuous function, such as the example in Figure 7.12. This shows the probability of a condition index being greater than a given value in relation to the age of the pavement. This type of function is sometimes known as a 'survivor curve'. The second type of function, known as 'markovian', contains dis-

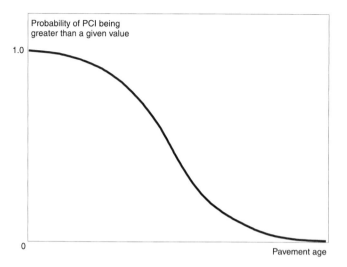

Figure 7.12 *Probability distribution (Source:* Shalin, 1994).

crete values: a range of possible conditions is divided into 'condition states'. At any point in time, probabilities are given for the likelihood of the pavement being in each condition state, and these are defined in a 'transition matrix'. This is then used to predict the probability of change in condition over time, typically yearly intervals. The following assumptions are made:

- The probability of making the transition from one state to another depends only on the present state.
- The transition process is stationary; the probability of changing from one state to another is independent of time.

An example of a simple probabilistic relationship is given in Table 7.3, in which the change in rut depth is predicted on the basis of the value in the present year. A more complex example is shown in Box 7.6. A further example has been given by Butt *et al.* (1994).

Table 7.3 *Example of a simple probabilistic condition prediction table for rutting*

Current rut depth (mm)	Probability (%) of rut depth being in this range in the following year						
	0	*1–4*	*5–9*	*10–14*	*15–19*	*≥20*	*Total*
0	93	4	2	1	0	0	100
1–4		90	6	3	1	0	100
5–9			85	11	2	2	100
10–14				86	11	3	100
15–19					80	20	100
≥20						100	100

Source: Kerali and Snaith (1992). Crown copyright 1992.
Reproduced by permission of the Controller of H. M. Stationery Office.

Box 7.6 — Example of probabilistic prediction model

This model estimates the future level of deterioration solely from the current condition. In this particular example, condition is characterised by a 'condition state' that takes its values from measurements of pavement roughness and surface distress, as shown in table below. This table utilises nine condition states, each indicated by a sequence number.

Roughness	Surface distress (per cent of area cracked)		
	0–3	3–7	>7
0–40	1	4	7
41–90	2	5	8
>90	3	6	9

The transition matrix in the table below was then set up to define the probability that a pavement in an initial state would deteriorate to a particular future condition state. This was based on formal interviews to capture the experience of practising engineers. Interviewees were asked to determine, to the best of their ability, the probability that a pavement in one of the condition states would progress to each of the future condition states in one time period. Normally, separate transition matrices are defined for each combination of factors that affects pavement performance.

Initial condition state	Probability of achieving future condition state								
	1	2	3	4	5	6	7	8	9
1	0.90	0.04	0.02	0.03	0.01	0	0	0	0
2	0.01	0.90	0.03	0	0.05	0.01	0	0	0
3	0	0.01	0.92	0	0.01	0.03	0	0.01	0.002
4	0	0	0	0.92	0.05	0.02	0	0.01	0
5	0	0	0	0.01	0.94	0.03	0.01	0.01	0
7	0	0	0	0	0.01	0.94	0	0.01	0.04
8	0	0	0	0	0	0	0.01	0.96	0.03
9	0	0	0	0	0	0.01	0	0.01	0.98

Adapted from: Haas *et al.* (1994).

7.6.3 Deterministic methods using mechanistic models

Mechanistic pavement design methods make tacit assumptions about changes in road deterioration over time. Most of these methods are based on the fundamental properties of a road pavement structure treated as a layered system. These depend on a knowledge of the constituent materials of the pavement (such as the Poisson ratio), an understanding of the parameters likely to be critical in assessing the life of the pavement (such as vertical subgrade strain, and tensile strain at the bottom of the bituminous layer), and a reliable method of determining pavement failure criteria against accepted norms. In truth, all of the methods are problematic but, nevertheless, are widely used. For instance, the *Shell* (1978) design method is an example of an approach that is based on fundamental mechanistic concepts. To be entirely credible, all of the available methods require empirical calibration, such as undertaken for trunk roads in the United Kingdom (Powell *et al.*, 1984).

Shahin (1994) considers that a purely mechanistic approach is applicable only to calculating a pavement response to external forces, such as traffic and climate. He considers that these methods should not be classified as predictive models, but that the calculated response, in terms of stress and strain, can be used as the independent variables in a regression prediction model.

7.6.4 Deterministic methods using regression models

Regression is the process where a dependent variable of observed or measured condition is related to one or more independent variables, such as traffic or axle load applications, pavement layer thicknesses and properties, subgrade strength, environmental factors and the like. Regression models are particularly applicable where a good historical database has been acquired.

Regression models can be of several different forms, including the following.
Simple regression:
$$y = a + bx_1$$
Multiple regression
$$y = a + bx_1 + cx_2 + dx_3 + \cdots$$
Polynomial regression
$$y = a + bx_1 + c(x_2)^2 + d(x_3)^3 + \cdots$$

By using transformations of variables in the regression process, relatively complex relationships can be derived.

Examples of the use of regression models for predicting pavement performance are given in Boxes 7.7 and 7.8. Further examples are given by Duffell and Pan (1996) and by Sadek *et al.* (1996).

7.6.5 Deterministic methods using mechanistic–empirical models

One of the difficulties with pavement deterioration data is that they tend to be very scattered, with the result that straightforward regression tends to produce relationships with low correlation coefficients. In addition, predictions using regression relationships that require extrapolation, rather than interpolation between the values used to derive the relationship, can be problematic. Indeed, sometimes predictions made with regression relationships produce results that are nonsense.

> **Box 7.7** *Example of a linear multiple regression model for predicting pavement deterioration*
>
> In this particular example, 25 years of data on roughness, surface distress, traffic, deflection, and other factors, have been used to develop deterioration models for granular pavements:
>
> $RCI = -5.998 + 6.870 \cdot \log_e RCI_B - 0.162 \cdot \log_e(AGE^2 + 1)$
> $\qquad + 0.185 \cdot AGE - 0.084 \cdot AGE \cdot \log_e RCI_B - 0.093 \cdot \Delta AGE$
>
> where
> $\qquad RCI$ = riding comfort index (scale 0 to 10) at any AGE in years
> $\qquad RCI_B$ = previous RCI
> $\qquad \Delta AGE$ = 4 years for the above equation, but can be any number of years as is appropriate.
>
> *Source*: Haas *et al.* (1994).

> **Box 7.8** *Example of a regression model using transformed variables for predicting pavement distress*
>
> The following equation predicts pavement deterioration in the form of an 'S'-curve:
>
> $D = e^{-(A/T)}$
>
> where
> $\qquad D$ = pavement distress (severity or extent)
> $\qquad T$ = time or traffic at which distress D is observed
> $\qquad A$ = parameter that represents the pavement characteristics.
>
> *Source*: Saraf and Majidzadeh (1992).

An approach to regression that can produce better results is to use a combination of mechanistic and empirical approaches. This enables an appropriate form of relationship to be established using knowledge of the mechanistic processes involved, and then to use empirical techniques, such as regression, to fit these relationships to available data. In other words, the analyst pre-determines the variables to be used and the general shape of the equation, and then uses regression techniques to determine the values of the coefficients in the equation (Kerali *et al.*, 1996).

Probably the best examples of such models are those incorporated into HDM-III (Watanatada *et al.*, 1987). A series of such relationships is provided for predicting future pavement deterioration for:

- Cracking initiation and progression
- Ravelling initiation and progression
- Pot-holing initiation and progression
- Surface damage at the end of the year before maintenance
- Rut depth progression
- Roughness progression.

> **Box 7.9**   *Mechanistic–empirical prediction of roughness*
>
> In this formulation, the change in roughness is predicted in terms of parameters reflecting:
>
> - The structural condition of the pavement, which is itself a function of pavement age, structural strength, cumulative axle loading, rutting, cracking and pot-holes
> - The initial value of roughness
>
> The relationship used is:
> $$\Delta R_d = 0.929 K_{gp} F + 0.023 K_{ge}(14 R_a - 0.714) + 0.714$$
> where
> ΔR_d = predicted incremental change in road roughness due to road deterioration during the year, in m/kmIRI
> R_a = roughness at start of the year, in m/kmIRI*
> K_{gp} = a user-specified deterioration factor for roughness progression (default = 1)
> K_{ge} = a user-specified deterioration factor for the environment-related annual fractional increase in roughness (default = 1)
> F = contribution to roughness of structural deformation and surface condition given by
> $$F = 134 EMT (SNCK + 1)^{-5.0} YE4 + 0.114(RDS_b - RDS_a) + 0.0066 \Delta CRX_d + 0.42 \Delta APOT_d$$
> where
> $EMT = \exp(0.023 K_{ge} AGE3)$
> $AGE3$ = the construction age, defined as the time since the last overlay, reconstruction or new construction activity, in years
> $SNCK$ = the modified structural number adjusted for the effect of cracking, and given by
> $SNCK = max(1.5; SNC - \Delta SNK)$
> SNC = the modified structural number, as defined earlier
> ΔSNK = the predicted reduction in structural number due to cracking since the last pavement reseal, overlay or reconstruction*
> $YE4$ = the number of equivalent standard axle loads for the analysis year, based on an axle load equivalency component of 4.0, in million/year
> RDS_b = the standard deviation of rut depth at the end of the year*
> RDS_a = the standard deviation of rut depth at the beginning of the year*
> ΔCRX_d = the predicted change in area of indexed cracking due to road deterioration*
> $\Delta APOT_d$ = the predicted change in total area of pot-holes during the analysis year due to road deterioration, in per cent.
>
> *Adapted from*: Paterson (1987) (see also this source reference for definitions of those variables marked with an asterisk*). ©The International Bank for Reconstruction and Development/The World Bank.

All relationships are deterministic, and an example of that for roughness progression is given in Box 7.9.

7.6.6 Calibration of deterministic methods to fit local conditions

Deterministic relationships are normally developed to predict typical conditions across a given range of pavement types. In all cases, they need to be calibrated to reflect local circumstances and conditions on individual sections of road. The principle of calibration is to *translate* and *rotate* the relationship in the following way:

$$y = a + bf(x)$$
where $f(x)$ = original predictive relationship
a = calibration factor that translates the relationship
b = calibration factor that rotates the relationship

Calibration in this way enables generic relationships to be modified to fit known data.

The concept of calibration can be developed further to enable relationships to be modified automatically, within a road management system, to take account of historical data collected over a number of years. Such an approach is clearly desirable.

A pragmatic approach to modelling pavement deterioration, which does utilise historical data, has been developed for the United Kingdom pavement management system. This utilises a very simple form of deterioration curve, but also has the ability to modify forward predictions based on the recent deterioration history of an individual road segment. It is described in Box 7.10.

Box 7.10 *Approach to fitting projection curves to historical data*

This approach works in the following way. For different road construction types, for different defects, and for each road hierarchy, condition is projected over time using simple curves of one of the following forms:

- Linear
- Power curve
- 'S'-shaped

The curves are user-definable and relate deteriorating condition to time. This reflects the normal situation where pavements are built with a structure that is appropriate to the traffic being carried, with the result that for a particular road hierarchy and construction type, pavements can normally be expected to deteriorate at similar general rates over time. Thus, separate relationships are set up for different:

- Defects
- Hierarchies
- Construction type

Rather than define these relationships as mathematical functions, the curves are defined as a series of discrete points, and intermediate values are determined automatically by linear interpolation. Each is defined as a relationship in two-dimensional space extending between (0,0) and (1,1). This relationship can be generalised using five constants (a, b, c, d and k), which define its range, as shown below.

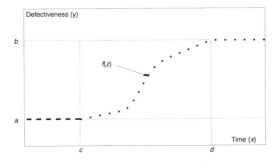

(continued)

Box 7.10 (contd.)

$$y = k[a + (b - a) f(z)]$$

where
$\quad z = (x - c) / (d - c)$
$\quad y = a$ when $x \leq c$
$\quad y = b$ when $x \geq d$
$\quad k$ = user-definable constant for factoring the whole relationship; normally $k = 1$.

The constants a, b, c, d define the co-ordinates of the limits of the curve for any particular purpose. By specifying $b < a$, the relationship gives decreasing values of y for increasing values of x. The form of $f(z)$ can be of any form specified by the user. By altering the value of k, special cases can be considered, such as pavements expected to deteriorate more quickly than the norm because of very high axle loading, or specific environmental aspects. As such, users can define their own standard curves and their own set of constants for each of the relationships. Over time, standard curves can be revised to reflect more accurately the performance of actual pavements in the network and to reflect different construction types, defects and hierarchies.

In order to fit curves to existing condition data, a simple 'shifting' (translation) and 'stretching' (rotation) process is adopted:

- The position of the standard deterioration curve is compared with the most recent data point; the curve is *shifted* along the 'time' axis until it passes through this point
- Where historical data exist, then the curve continues to be fixed to pass through the most recent point, but is *stretched* or shrunk along the 'time' axis to fit historical data points as well as possible; least squares fitting is used for this, giving more weight to the most recent data points
- Condition at future points in time can then be predicted using the re-positioned curve

It has been found that, notwithstanding the variability of condition data, by a sensible choice of curve shape, and by use of the shifting and stretching process, good projections of future condition can be obtained. The method reflects more accurately the peculiarities of individual sections by incorporating this knowledge into the modified projections.

Adapted from: Phillips (1994).

7.7 Observations on prioritisation methods

7.7.1 Key features of available methods

It has been reported (Phillips, 1994) that the use of condition-based prioritisation ('worst first') methods to allocate funds when budgets are tightly constrained leads inevitably, over time, to budget needs increasing and to road conditions deteriorating. Similar conclusions are also reported by Haas et al. (1994). However, in such budget situations, third generation methods are often able to maintain or even improve road conditions.

Phillips also reported that, in simple terms, treatments should be applied at the latest possible time before deterioration dictates that a more expensive treatment would be required. For example, an overlay should be deferred until just before the point at which a pavement reconstruction would be triggered; applying the overlay before this time would mean that the full value was not obtained from the existing pavement; failing to apply the overlay on time would result in a large increase in treatment cost. The actual decision-making process also needs to take into consideration many factors other than this, and certainly should include the effects on road user costs.

The approach to prioritisation, therefore, needs to take account of the following:

- Where virement of funds is not permitted, prioritisation must be undertaken separately for each budget head; this will normally limit the treatment choices to be included in any particular evaluation.
- Intervention levels should be set at a maximum level such that treatments are triggered just before a more expensive treatment is needed, with an allowance for error and uncertainty.
- Priority should, in general, be given to cheaper preventive treatments where these prevent the need for more expensive treatments in subsequent years.

However, there will also be a need to prioritise the minor activities, such as kerb works, road lining, road furniture replacement and renewal, within the same list. Many of these activities are time-dependent, and their schedules need to be combined with the above prioritisation process. Decisions on this need to reflect local practice.

7.7.2 Data requirements

Moving from first through to third generation methods increases the need for data. Thus, any desire to use the more sophisticated methods must be tempered by the demands that this might place on data collection resources. These issues were discussed in Chapter 5.

A large road administration, with a reasonable level of funding and good skill resources, should be aiming to utilise third generation methods. However, a small administration, perhaps based in a municipality, may need to adopt a simpler approach. In such cases, the use of treatment-based methods (sub-section 7.2) can overcome many of the problems of using a 'worst first' approach, and should be considered in such cases.

7.7.3 Use of third generation methods

It is interesting that several developments in the area of management systems over the past few years have all come up independently with variations on the approach of using the economic boundary (Transportation Research Board, 1994). With this approach, the drawbacks identified in section 7.5 still apply. These were in the areas of:

- Uncertainty of traffic projections for long periods into the future
- Problems of traffic congestion in later years of the analysis
- Need to forecast the effects and durability of maintenance treatments for long periods into the future

- Assumptions about the economic justification of future treatments
- Assumptions about future funding levels.

It is for this reason that the methods were introduced for analysis over shorter time periods than the full life cycle. The methods treat future costs in such a way as to reduce computational effort to satisfactory limits while, at the same time, producing results that are close to the 'optimum'.

It could be argued that only at the planning level is there a need to set priorities and to optimise over multiple years. In this case, full life cycle analysis may be the most appropriate method. But the penalty of this choice is the significant computational requirement. Thus, if life cycle analysis is used in conjunction with priority optimisation for planning, only very few variables should normally be considered. Consideration should also be given to using the year-by-year analysis method discussed earlier.

For undertaking the management function of programming, the need is normally to decide how best to spend budget allocations in the relatively short term. Reduced time period analysis methods can reduce computational requirements significantly in this situation. Since the budgeting process is normally repeated each year, long-term decisions are continually adjusted over time, so sub-optimal decisions are corrected.

8 Operations Management

- Routine maintenance
  - management tasks and work types
  - cyclic and reactive works
  - scheduling
- Winter maintenance
  - defining aims
  - assessing needs
  - determining actions
  - costs and priorities
- Choice of technology
  - technological appropriateness
  - cost-effectiveness
  - availability of labour and equipment
  - domestic resources
- Equipment management
  - autonomous management
  - organisation of maintenance and repair
  - administrative and commercial management
- Monitoring
  - key indicators of effectiveness
  - value for money

8.1 Operations considered

The operations phase is the cutting edge of road maintenance. However well activities are planned, programmed or prepared, the product that is delivered finally on the road will depend on how well operations are managed. Effective management helps to ensure that work is undertaken efficiently to meet requirements of quality, time and budget.

Some of the means of achieving these requirements have already been discussed. For example, basic management issues, including quality management, were discussed in Chapter 1; the achievement of effectiveness and efficiency was discussed in Chapter 2; methods of cost control were described in Chapter 4; and the use of relevant data was considered in Chapter 5. Other issues, such as project management, are beyond the scope of this book, and reference should be made to one of the many texts on the subject, such as that by Freeman-Bell and Balkwill (1993).

This chapter deals with the management of individual activities. It focuses particularly on the more difficult management areas of routine and winter maintenance. Periodic maintenance and development activities are not included, since the techniques for these are essentially the same as those for new construction and are well covered elsewhere (for example, by Millard, 1993). Also, no attempt is made here to provide detailed descriptions of the techniques of road maintenance. Again, there are a number of texts that cover this, including those by TRRL Overseas Unit (1985), PIARC (1994) and Atkinson (1997). The chapter goes on to consider issues concerning the choice of technology and the appropriate use of labour-intensive and intermediate technology methods. The main issues affecting the management of equipment are then described. Finally, the process of monitoring is considered.

8.2 Routine maintenance

8.2.1 Management tasks and work types

The management of routine maintenance can be considered to include the following tasks (Atkinson, 1997):

- To examine, appraise and validate local conditions
- To prepare realistic works programmes in order of priority
- To determine levels of funding required
- To prepare arguments that justify proposed works and costing, together with the consequences of alternative courses of action
- To ensure that actions taken demonstrate the advantages claimed
- To improve the gathering, recording and use of available data concerning all areas of responsibility.

The nature and requirements for routine maintenance will vary depending on the ownership and importance of individual roads, and on the features being maintained. In addition, requirements will depend on the detailed contents of the policy framework's objectives, standards and intervention levels, as described in Chapter 2.

As noted in Chapter 1, it is convenient from a management point of view to consider routine maintenance under the two headings of cyclic and reactive works. This is because the planning and management of these work types are fundamentally different.

8.2.2 Cyclic works

These are scheduled works whose needs are dependent on environmental effects rather than on traffic, although scheduling of works is being used increasingly for a wide range of activities because it simplifies planning, programming and budget formulation. Cyclic works include such activities as road cleaning, drainage maintenance and vegetation control. These are all activities that can have a relatively high profile from a public and political point of view.

The frequency of cyclic works should be based on historical experience. Reference to the policy framework should assist with determining frequencies by defining standards and intervention levels for activities that will assist in meeting policy objectives. It is therefore important that appropriate standards and intervention levels are set to be consistent with the budget that is likely to be available for the various activities.

Programming cyclic works can only be undertaken in a satisfactory manner within budget constraints where there is a detailed item inventory for the road network (see

Chapter 5) and good unit cost information (see Chapter 4). The required budget for any particular cyclic activity is the product of the quantity of the relevant inventory item, the number of times a year that the activity will be undertaken, and its unit cost. However, the cost of collecting inventory information, particularly on minor roads, must be balanced against the benefits that can be obtained from better management. Examples of savings that have been obtained by collecting appropriate management information are given in Box 8.1. The appropriate level of detail of inventory collection should be tested by a cost–benefit analysis.

Box 8.1   *Examples of cost savings for cyclic maintenance resulting from the use of good management information*

Grass cutting

Problem:	No clear policy on standards leading to patchy service levels; insufficient information to compile unit costs; budgeting could be nothing but historically based
Action:	Inventory prepared covering all areas of the network; cyclic maintenance frequencies adopted, based on national guidelines, modified for local conditions; intervention levels defined
Outcome:	Budgets prepared on an assessment of needs based on reliable unit cost; work subject to competition and carried out to consistent standards throughout the road administration; annual cost reduced by £121 000 ($180 000)

Checking of street lighting energy accounts

Problem:	The electricity supplier invoiced the road administration based on historic information about the number of lamps, the type of lamps and switch gear, and the burning regime (all-night or part-night lighting); bills were paid as rendered because no better information was available with which to check them
Action:	Street light maintenance was put out to open tender and the successful tenderer was required to collect full inventory information as part of the contract; the inventory was then compared with the electricity supplier's charging schedules
Outcome:	The total number of lamps being charged was 9800 more than remained in service; the types of lamp being charged for were at variance with those in service, and the conversions to energy efficient lamps and control gear had not been reflected in the invoices; the administration negotiated a repayment from the electricity supplier, and there are on-going savings of £125 000 ($190 000)
Adapted from:	Audit Commission (1988). Crown copyright 1988. Reproduced by permission of the Controller of H. M. Stationery Office.

Database management software is available to assist with managing cyclic maintenance activities using information stored about the road network and its item inventory. Such software can typically undertake the following functions:

- Perform quantity calculations on the network and inventory data, typically of counts, lengths or areas.
- Multiply the results of the calculations by a maintenance frequency and unit rate.
- Report on the quantities and costs accumulated by various criteria, determined from network or inventory data, where
  - multiple types of inventory calculations may be reported upon
  - frequencies and rates may be defined at various levels, from network-wide defaults down to specific values for individual items of inventory
  - different cyclic activities on the same section of road are undertaken by different road administrations (for example, the signs on a minor road may 'belong' to the owner of a main road with which it intersects)
  - different budget scenarios can be investigated.

8.2.3 Reactive works

These are works that are carried out to respond to minor defects caused by a combination of traffic and environmental effects. Reactive maintenance works are likely to result from:

- Routine inspection and assessment procedures
- Road accident reports
- Complaints from the public.

The following types of inspection are in common use:

- *Safety inspections*
  Required to identify where immediate, urgent or emergency works are needed
  - immediate works include activities such as removing dangerous debris and the placing of temporary warning signs, and are undertaken by the inspection teams themselves
  - urgent and emergency works are those that must be undertaken within a short timescale (typically less than 24 hours) to repair a dangerous defect or a cut road, respectively.
- *Detailed inspections*
  Required to identify where reactive maintenance works are needed, although a safety inspection would normally be carried out at the same time.
- *Special inspections*
  An unscheduled inspection required where a defect has been brought to the attention of the road administration as the result of an accident report, a complaint from the public, or by notification from some other source.
- *Monitoring inspections*
  Undertaken to provide a check that work has been carried out to agreed requirements (technical audit), or to feed back experience of actual performance into the management process (monitoring).

Whereas inspections for periodic works identify the need for maintenance to be programmed in the subsequent budget period, routine inspections identify reactive works that need to be scheduled into the current year's programme. On the more heavily trafficked roads, detailed inspections may be carried out several times a year. On motorways, for example, safety inspections may even be carried out daily. Inspections can also provide the basis of statistical comparisons and technical audit of work already undertaken for budgetary control and resource management purposes, or for the comparison of performance against standards and intervention levels. Inspection frequencies should be defined as part of the policy framework to ensure consistency across the network, but recognising that frequencies may vary depending upon road hierarchy. Standards and intervention levels for the reactive works identified by the surveys may also vary in a similar manner. Thus a system of inspections, standards and intervention levels form the basis of effective management in this area.

However, there are several differences between the management of reactive works and those, say, of periodic works. These pose particular problems for the maintenance manager. Periodic works have much longer durations and are generally easier to define than reactive works. As a result, treatment selection and prioritisation methods, such as those described in Chapters 6 and 7, are relatively easy to apply. The use of such methods is not normally realistic for reactive works, and less formal methods need to be used with the consequence of sub-optimal works being undertaken when budgets are constrained. However, standard procedures for prioritisation should still be used that are based on experience, although it is best to avoid relying *solely* on the subjective judgement of individual local engineers.

The difficulty of specifying reactive work requirements means that defining appropriate standards and intervention levels is more difficult. The cost of surveys to identify reactive work requirements is a much higher proportion of the cost of the works than for periodic maintenance, calling into question the viability of undertaking surveys in all situations. The short duration of activities means that operational requirements are constantly changing. This is exacerbated by large fluctuations in demand for works because of climatic and other effects beyond the control of the manager. These issues result in the requirement for high levels of management expertise, supported by reliable and up-to-date management information, to ensure that a tight control of operations is maintained.

Computer systems are available for controlling the day-to-day operational requirements associated with maintenance operations. These, typically, are based around the use of performance standards for works activities, as described in Chapter 4. Such systems can assist with the issuing of work instructions and recording work accomplishment, and can produce useful summaries of productivity and resource utilisation at appropriate intervals of time. Other systems are also available for assisting with the management of routine maintenance, and an example is given in Box 8.2. This system manages routine inspections. It also identifies where repairs of defects have not been undertaken since the previous inspection and are, therefore, overdue. Works orders can be issued and audit reports produced.

8.2.4 Scheduling

The responsibility for managing routine maintenance operations will depend on who is responsible for undertaking the various activities: the road administration itself, maintenance management consultants, works contractors, or a mixture of all three. Different methods of procuring maintenance services are discussed in Chapter 9. A key management task in this area is the scheduling of works. A maintenance schedule can be considered as a diary of works. A works diary provides a basic management tool for both scheduling future works and for recording past works. It will need to be updated frequently and, in many situations, the logging of events will be required every day. The diary can include both cyclic and reactive works, and the various inspections that need to be undertaken, where these are the responsibility of the same organisation. It also provides a useful source of information when dealing with enquiries and complaints from the public and elected representatives.

Box 8.2   *Routine maintenance management system*

The UK Department of Transport Routine Maintenance Management System (RMMS) has the following features:

- Identification of the required cyclic maintenance and inspection activities based on a section's recorded inventory information
- Recording of inspections and the defects found, allowing direct entry into a data logger
- Classification of defects to give different repair interval requirements
- Automatic or manual matching of defects found in inspections with those already existing on the network
- Creation of works orders for both cyclic and reactive maintenance

Reports obtainable from RMMS include:

- Schedules of activities required
- Cyclic maintenance, inspection and defect listings
- Numbers of defects and associated quantities, such as total length and area
- Audit reports providing a statistical comparison of performance against standards.

Irrespective of where the responsibilities lie, the essential process of scheduling is the same. Cyclic, reactive works and inspections can be treated as discrete 'projects', each with a start and an end, and with a duration and resource requirements. The scheduling problem of each of these projects can then be treated as a conventional network analysis that can be solved using critical path techniques (Freeman-Bell and Balkwill, 1993). Thus, scheduling can be considered as a project management task (see Chapter 1), and proprietary project management software can be used to assist with this. Such a system can be used in conjunction with a works diary.

8.2.5 Utility works

Particularly in urban areas, roads are subject to works undertaken by statutory authorities in the course of laying and maintaining their pipes and cables. Such works can result in considerable damage to the fabric of the road. The scale of the problem is illustrated in Box 8.3.

Box 8.3	Public utility activity in a typical year in the United Kingdom		
Utility	New and replacement mains (km)	New and replacement services	Small openings
Electricity	4 000	200 000	217 000
Gas	5 457	767 000	544 000
Telecomms	3 150	467 000	74 000
Water	3 200	230 000	536 000
Sewers	2 500	—	—
	18 307	1 664 000	1 371 000

Source: Atkinson (1997).

Responsibility for authorising, supervising and monitoring utility works depends on the detailed legislation in individual countries, and it is not possible to generalise in this area. However, in many cases the management of these activities is a major problem for road administrations. The criterion that 'whoever damages the road should be responsible for its reinstatement to an agreed standard' is a sound basis for the organisation and management of utility works. This is similar to the concept of 'the polluter pays' used in other sectors. However, such an approach does need to be supported by legislation, standards and codes of practice, as well as by the provision of resources for inspection and sanction by the road administration.

8.2.6 Key considerations

Some key areas for consideration are given in Box 8.4.

8.3 Winter maintenance

8.3.1 The winter problem

In many countries, winter conditions cause disruption of road traffic. The magnitude of this varies with geographical location, altitude and other climatological factors, such as whether the climate is inland or maritime. In most areas, the severity of winter conditions varies from year to year. Modern society has become more vulnerable to these effects because of the disruption to traffic that is caused: factories now have relatively small stores and depend on 'just-in-time' deliveries; timetables for buses are the same in summer and in winter, and society is generally more mobile, relying on good road conditions when planning journeys. Winter maintenance cannot solve all problems that arise in this area. Road users need to play their part by adjusting their way of driving, and by being properly equipped, for example with winter tyres.

Swedish investigations have shown that the accident risk increases five to ten times on main roads with 30 mm of snow compared with dry roads. Accident reports to insurance companies suggest that a road covered by slush can have up to 50 times the accident risk compared with a dry road. Accident risks are also higher in areas subject to only intermittent snowfalls compared with those where heavy snow is commonplace. The adverse impacts on speed and vehicle operating costs decrease quickly when snow stops falling, as shown in Table 8.1.

Box 8.4 *Key areas for consideration in routine maintenance*

Inventory
- Is there an adequate and up-to-date inventory?
- Does it include the inventory of agents who carry out works on behalf of the road administration?

Standards and frequencies of intervention
- Has the policy framework in respect of standards and frequencies been documented?
- Is a clear distinction made between works carried out for amenity/environmental purposes and safety/repair purposes?

Budgeting
- Is the budget calculated according to the formula:
    inventory × unit cost × frequency?

Client–supplier split
- Is there a separation within the administration between the client and the supplier role; for example, are the maintenance and inspection functions for patching works and street lighting separated?
- Is there client-side supervision of the works organisation?

Tendering
- Is the majority of routine maintenance put out to tender?
- In particular
    – is the road administration dependent on the electricity supplier for street lighting maintenance?
    – has traffic signal maintenance always been done by the equipment supplier?
    – if road markings are maintained by the administration's own works unit, what is the justification for this?

Supervision
- Is the level of client-side supervision adequate?
- Are the administration's own works unit and contractors' internal supervision levels adequate?
- When were the administration's works unit's bonus schemes last reviewed?

Adapted from: Audit Commission (1988). Crown copyright 1988. Reproduced by permission of the Controller of H. M. Stationery Office.

Compared with most other maintenance activities, winter maintenance is characterised by the need for a very fast response, and by the impacts of the measures taken being short-lived. Each road administration needs to create an organisation capable of providing a degree of security for traffic operations during times of adverse weather that is appropriate to local circumstances.

Table 8.1 *Impact of winter conditions on vehicles*

	30 mm of snow			60 mm of snow		
	First vehicle	*100th vehicle*	*1000th vehicle*	*First vehicle*	*100th vehicle*	*1000th vehicle*
Accident increase	4–8 times			6–12 times		
Speed reduction (%)	10–15	5–10	0–10	20–30	10–20	5–10
Increase in vehicle operating costs (%)	2–8	0–5	zero	5–15	0–5	zero

8.3.2 The winter maintenance management cycle

When determining its policy framework (see Chapter 2), the road administration needs to define objectives, standards and intervention levels for winter maintenance. It also needs to devise a strategy for achieving these. This needs to be detailed in the policy document of the administration, and should cover, for example, the priorities for action within the road network, the response times, snow clearing and salting routes, salt spreading rates, and the personnel strategy (Audit Commission, 1988). The management cycle (see Chapter 1) provides a framework that can be used to assist with implementing this strategy. Particular actions or decisions required to be taken under each of the steps of the management cycle are:

1. Define aims
   - priority routes
   - response times
2. Assess needs
   - snow clearing
   - de-icing
3. Determine actions
   - snow clearing and salting routes
   - call-out criteria
4. Determine costs and priorities
   - resource requirements
   - budget needs
5. Implement activities
   - in-house units
   - contractors
6. Monitoring and audit
   - monitor performance
   - review and refine aims

Issues arising in the first four of these steps are discussed in the following sub-sections. Implementation issues involving the use of in-house units and contractors are discussed in Chapter 9, although a detailed description of techniques is beyond the scope of this book (see Atkinson, 1997); monitoring and audit is discussed in section 8.6.

8.3.3 Defining aims

Standards and intervention levels for winter maintenance are established as part of the policy formulation process. It is normal to define the priority between the different routes in the road network for salting and snow clearing, recognising that this needs to reflect road hierarchy, traffic flow and the different demands for winter maintenance. Requirements for continuity of activities, available funds and special conditions may necessitate divergence from the normal standards and intervention levels.

Road administrations may wish to define salting and snow clearing targets. These can be set in terms of the maximum time between the decision to salt or clear and the vehicles' return to the depot having completed the work. Typical targets are two to three hours, although the range can be between one and six hours (Local Authority Associations, 1989). Targets can also be set separately for the response time to start salting and the time for undertaking the work. Some administrations set additional aims, such that roads must be treated before 7.30 am so that roads are clear before the morning rush hour. Priorities normally reflect the general road hierarchy (see Chapter 2), but high priority is often attached to roads leading to important establishments, such as hospitals, bus routes, major commuter routes and the like. It is important that consistent standards and intervention levels are adopted.

The various Nordic countries, where winter maintenance is a major activity, have adopted standards that have many similarities. The standards determine the level of service that is provided to each road. Roads are classified according to their functional hierarchy and traffic volume, as in Table 8.2. The different classifications each have associated with them different standards and intervention levels.

Table 8.2 *Example of winter maintenance route classification used in the Nordic countries*

Traffic (ADT)	Trunk roads	Main roads	Local roads
> 12 000	A	A	—
6000–12 000	A	B	C
2000–6000	B	C	C
500–2000	C	C	D
< 500	—	D	E

Note: A–E reflect different winter maintenance standards and intervention levels.

8.3.4 Assessing needs

Snow clearance

Requirements for this should be related to the number of times that it snows and the degree of difficulty in clearing the snow. The degree of difficulty depends on:

- Intensity of the snow fall
- Duration of the snow fall
- Wind
- Temperature of the road and air
- Topography.

The relationship between the number of snow falls and the degree of difficulty of clearing is shown schematically in Figure 8.1. Clearly, some snow falls are so light that there is no need for a response. Conversely, it is unrealistic to expect a road administration to be able to respond in all situations. In extreme situations, users have to accept delays. Figure 8.1 also illustrates the ranges of values where action would normally be undertaken. For the highest class of road (Nordic Class A), only a very few snow falls are so light that there is no need for clearance; whereas on the lowest class (Nordic Class E), users have to accept considerably worse conditions before snow clearance will start. When conditions are severe, resources may be moved from lower to higher class roads to keep these open for as long as possible.

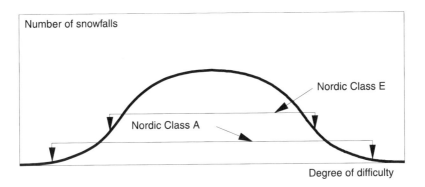

Figure 8.1 *Schematic representation of the relationship between the number of snowfalls and the degree of difficulty of clearance*

Depending on the geographic area concerned, the shape of the snowfall/degree of difficulty curve will alter, as illustrated in Figure 8.2. The position of the curve can suggest the type of organisation necessary to deal with the winter maintenance problem. For example, the organisation could be the same in both the arctic and sub-arctic situations, but the fixed cost (the 'insurance' cost) would be considerably higher in the sub-arctic case; in the alpine climatic region, the use of different types of equipment such as snow-blowers would probably be justified; in the temperate case, it would be expensive to keep an extensive snow-removal organisation in a state of preparedness to deal with only a few snowfalls, and it may be more economical to wait for the snow to melt. Decisions on the appropriate level for the fixed insurance cost should be taken by considering the impact and cost on road users during snowfalls.

Standards for snow clearance are normally determined through a relationship between snow depth and response time, such as shown in Figure 8.3 for the Nordic countries. Standards can either be defined as the amount of snow allowed on the road during snowfall, or by the acceptable cycle time for the snow clearance.

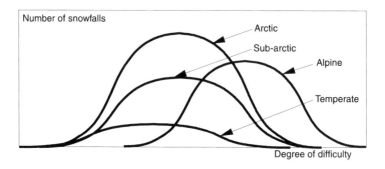

Figure 8.2 *Schematic representation of the snowfall /clearance difficulty relationship in different climatic regions*

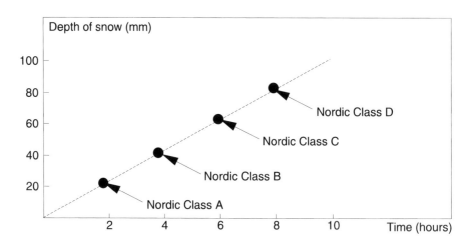

Figure 8.3 *Relationship between allowable snow depth and response time for Nordic countries*

De-icing

This is used to reduce the slipperiness of the road surface. It is also known as 'salting' or 'gritting', reflecting the use of different de-icing materials. Slipperiness can be caused in a number of ways:
- Freezing
  - wet surface freezes
  - hoar frost causing black ice

- Precipitation and fog
  - rain onto a cold surface
  - freezing rain
  - snow
  - freezing fog.

Control of ice can be undertaken both chemically and mechanically. Salt (NaCl) is typically used to keep roads free from ice and snow when traffic levels are greater than 1500–2000 vehicles/day. On lower volume roads, chemically-inert grit is normally used, and a snow-packed surface may have to be accepted. However, grit is quickly removed from the road by traffic and, where sand is used, this can be removed by as few as 150 vehicle passes. As a result, sand is normally only used when it is very slippery, and then mostly on critical stretches such as intersections, curves and steep hills. A bigger difference in level of service normally has to be accepted by drivers on minor roads compared with major roads, as shown in Figure 8.4. Here the difference in level of service on major and minor roads for snow clearance (as shown in Figure 8.1) will be seen to be much less.

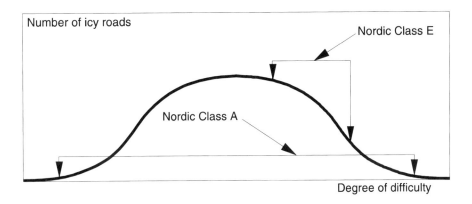

Figure 8.4 *Schematic relationship between icy roads and the degree of difficulty in removal*

The use of salt is damaging to the environment, and also to concrete roads and bridges. It leads to complaints from the public. As a result, ways are being sought to reduce the amount of salt that is used. Trials have been undertaken in the Nordic countries to investigate the impact of reduced salting on traffic accidents. The results of the *Minsalt* trials in Sweden are summarised in Figure 8.5. These found that there was some uncertainty in the effect of reduced salting on safety, especially on roads with traffic of around 2000 vehicles/day. In addition, the decrease of slippery conditions over time has not always resulted in the expected reductions in road accidents. A reason for this may be that drivers find it easier to detect if a road is slippery when salt is not used. Another reason may be that drivers do not reduce speeds enough on roads that they believe to have been salted.

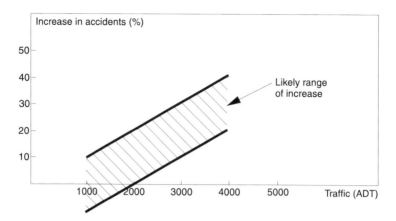

Figure 8.5 *Increase in accident rate through not salting roads*

New methods are also being used to reduce the use of salt. In the past, the use of dry salt was normal, but it was found that this tends to fly away from the road surface before the ice starts to melt. Pre-wetted salt or brine are now more commonly used to overcome this problem, and have the advantage that smaller quantities can be applied. Salting as a preventive measure also has benefits because it takes less salt to prevent water from freezing than to melt ice. Bridge decks need special consideration because they ice-up more quickly than adjacent road pavements. Weather information systems used together with weather radar, thermal mapping and satellites enable risk situations to be predicted with reasonable reliability. Preventative salting stops roads from becoming slippery and results in a reduction in accidents compared with reactive methods, as shown in Figure 8.6.

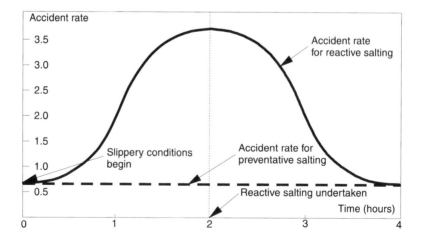

Figure 8.6 *Accident reduction resulting from preventative salting*

8.3.5 Determining actions

There are significant savings to be made by rationalising the clearing and de-icing routes to reduce non-productive travel. In the United Kingdom, the most efficient salting routes tend to have a length of about 50 kilometres, which is consistent with the typical response time of two to three hours favoured by road administrations (Audit Commission, 1988). Routes should be planned centrally, and should run over area boundaries where this is more efficient. Use of modern salt spreaders means that both sides of the road can be salted in a single pass, eliminating the need to drive along both sides of the road. In addition to saving time, the amount of salt used is also often reduced.

A call-out of salt spreaders in a road administration can be an expensive undertaking. (The Audit Commission estimates that, for a typical county in the United Kingdom, this figure is in excess of $20 000 for a single mobilisation.) On the other hand, it is argued that costs to road users can be very high if the road becomes slippery and nothing is done. It is, therefore, vital that decisions are based on good information. The availability and quality of information available to winter maintenance organisations have improved considerably in recent years through both better access to meteorological forecasts, and by increased reliability and use of road-side sensors.

Weather information instruments along roads can measure air humidity, and both air and road surface temperature continuously. The equipment can also measure wind direction and speed, and also the intensity of precipitation. Instruments can be linked to computer systems that can forecast with good reliability the risk of ice forming on the road surface. Sensors need to be located at critical points in the network known as 'cold spots'. The required number and positions of these can be determined by 'thermal mapping' of the road network. Thermal mapping can also assist the design of clearing and salting routes or restrict the scope of call-outs in marginal conditions. However, such an approach requires a high initial investment, both in terms of the price of sensors and the telecommunication networks and computer systems needed to support them, and in terms of the price of the thermal mapping itself. Hence, a cost–benefit analysis is needed prior to investing in this technology.

In many cases, changes in the weather can be monitored using satellite and weather radar. Even so, there is considerable potential for improving local forecasting methods, particularly for the onset, duration and intensity of snowfall.

8.3.6 Determining costs and priorities

The policy framework in the area of winter maintenance should be used to set budgets and priorities. Inevitably, because of the unpredictable need from year to year, the budgeting process needs to be based on historical precedent, perhaps with past costs considered over at least a five year period. Budgets should take into account:

- Route length and the priorities between routes
- Number of days when frost can be expected

- Preferred spread rates for salt or sand
- Number of days when snow can be expected
- Clearance rates for snow.

It is often convenient to have access to contingency funds that can be called upon in the event of unusually prolonged or severe winter conditions.

Winter maintenance can be undertaken by in-house units or by contractors, or by a mixture of the two. There is some scope for the use of small contractors and farmers for snow clearing, particularly in rural areas. General issues about the procurement of road maintenance works are discussed in Chapter 9, but some special considerations for winter maintenance are considered here. Some comments on staffing issues are given in Box 8.5.

Box 8.5 *Issues concerning the use of labour for winter maintenance*

There are four main strategies for out-of-hours winter maintenance operations:

- Stand-by
- Stand-to
- Full night shift
- Voluntary system

Most road administrations operate a stand-by system during the winter season, calling out staff, who normally do other work, as needed. In most cases, the staff receive a stand-by payment, irrespective of whether they are called out, and are paid overtime for actual call-outs. Stand-to systems involve no stand-by payment, but staff report typically at 05.30 each morning, irrespective of weather conditions, and are paid whether they turn out or not. Some administrations operate permanent night shifts. These are expensive and are normally justified only in rural areas with long salting and clearing routes, or where the response times would be excessive if crews had to be called from home. Voluntary systems involve no payments except when called out, and rely heavily on the goodwill of staff.

Stand-by periods relate to the anticipated length of the winter maintenance season, with obvious repercussions about the amount paid in stand-by allowances. There will often be scope for reviewing the length of periods as a result of monitoring records of actual weather conditions and turn-outs to ensure that appropriate and cost-effective values are being used. The number of staff on stand-by depends on the number and priority of routes and, again, should be subject to monitoring and review. As a broad rule, an average of 1.5 staff including supervisors, mechanics and depot hands per route is needed if machines have single operators; the equivalent average is 2.4 per route if two operators are used. While double operation is usual for snow clearing, it is not needed for most salting operations.

Incentive bonuses, if paid at all, should be a pre-determined fixed percentage, or linked to the average paid for other road works in the relevant period. Bonus schemes that are linked to speed over the route or volume of salt spread simply encourage operators to exceed optimum speeds for spreading and to spread at much higher rates than necessary.

Adapted from: Audit Commission (1988). Crown copyright 1988.
   Reproduced by permission of the Controller of H. M. Stationery Office.

Issues concerning the use of pre-wetted salt and brine have already been discussed. However, all salt-based materials are highly corrosive and environmentally damaging.

The total cost of using salt needs to be considered in the light of the cost of damage to the road structure, bridges, utilities, vehicles and the environment more generally. Other materials, such as urea, liquid polyglycol and calcium magnesium acetate, can be used, but are expensive and their use can normally only be justified on bridges or long sections of elevated road. Given the continuing need to use salt, consideration needs to be given to its storage, since good practice can reduce losses through leaching, and result in more efficient spreading by avoiding contamination and other problems. The use of simple covered storage facilities can result in significant savings (Audit Commission, 1988).

8.4 Choice of technology

8.4.1 Technology options

Some maintenance activities are suitable for carrying out by machinery, whereas others are more appropriate by manual methods. However, many activities offer the option of either method. When considering the most appropriate technology, it is not necessarily a simple choice between labour and equipment-intensive methods: for some activities, there is sometimes an intermediate solution which utilises significant amounts of labour in conjunction with simple machinery for certain operations. For example, regravelling can be undertaken by both labour and equipment-intensive methods, but can also be done using labour for winning, loading and spreading gravel but, where haul distances are significant, using tractors and trailers for the haulage.

Thus, the three choices of technology that will be considered are:

- *Labour-intensive*
  Use of labour for all activities, utilising only unpowered hand tools, and walking to site from home or road camp.
- *Intermediate*
  Labour-intensive, as above, but supplemented by tractor-based transport and simple equipment, and transport to site where appropriate.
- *Equipment-based*
  Predominant use of equipment with high output.

The phrase 'labour-based' is also sometimes used, and refers normally to what is defined here as 'intermediate'.

It was noted in Chapter 2 that increasing specificity and competition of maintenance operations improved the effectiveness and efficiency. It should be noted that all of the technologies can be organised in such a way to achieve both specificity and competition, and that there is no intrinsic difference in the suitability of different technologies in terms of meeting these aims. However, some criteria for choice of technology that do need to be considered are:

- The ability of the technology to meet the designated standards, recognising that these may vary depending upon
  - road class and traffic level
  - geographic and climatic conditions.

- The cost-effectiveness of the approach, recognising that this may depend upon
  - the relative cost of labour and equipment
  - the procedures adopted and the effectiveness of management of the operation.
- The availability of both labour and equipment, recognising that this may depend upon the demographic and socio-economic conditions in different parts of the country for labour, and upon the type of equipment being considered.
- A preference for many countries to use domestic resources rather than those that are imported.

Technology options for maintenance activities are reviewed below against these four criteria.

8.4.2 Technological appropriateness

Technological choice is of most concern when dealing with unpaved roads and off-road maintenance activities. Table 8.3 gives examples of these activities, and notes the different technological options that are appropriate for carrying them out.

8.4.3 Cost-effectiveness

Studies carried out by the International Labour Office (ILO) suggest that it is more economic to use labour-intensive methods where real labour costs are less than about US$4–6 per day. Such a break-even figure clearly rules out these methods for use in industrialised countries, but many developing countries have labour costs that are lower than this figure, making such methods attractive from an economic point of view. Countries with emerging economies may also meet this criterion: real labour costs need to be considered in each particular case. The following only applies to those countries where labour rates meet this criterion.

Experience from work undertaken in Kenya on the *Rural Access Roads Programme* and the *Minor Roads Programme* indicates that tractor–trailer haulage is cost-effective for haul distances up to about 10 kilometres, and that truck-based haulage should be used for distances greater than this. Where tractor-based haulage is used, and several trailers are available for each tractor which can be loaded alternately, then loading of material can be undertaken by hand.

A further programme in Kenya, known as *Roads 2000*, recommends that, because of vehicle operating cost considerations, labour-intensive surface patching operations on unpaved roads are only justified when traffic levels are below about 50 vehicles per day; above this traffic level, intermediate technology reshaping using towed graders is necessary. This conclusion would also apply to the spreading of material by hand in connection with spot or full regravelling operations. The report on the *Roads 2000* programme also compares the costs of partial rehabilitation using the three possible choices of technology. The results are summarised in Table 8.4, which shows that the intermediate technology approach, using labour supplemented by tractors and towed graders, is the cheapest approach for this activity in this particular example.

Table 8.3 *Appropriateness of maintenance technology for different activities*

Maintenance activity	Methods which are suitable		
	Labour-intensive	Intermediate	Equipment-based
Patching of unpaved roads	Essentially a hand operation, but requires provision of patching material	Appropriate for winning and haulage[1] of patching material, and compaction	Appropriate for winning and haulage of patching material, and compaction
Grading unpaved roads	Hand methods can be used, but cannot achieve required finish where vehicle operating costs are significant	Appropriate – using towed graders	Appropriate – using graders (plus compaction)
Full or spot regravelling of unpaved roads	Spreading can be undertaken by hand, but requires provision of gravel; see also comments under grading	Appropriate for haulage[1] of gravel; towed graders could be used for spreading, depending on length of works	Appropriate for winning, haulage and spreading of material
Maintenance of side drains	Appropriate using lengthmen or labour gangs (transport may be needed for gangs)	Appropriate using towed graders for V-shaped ditches; no suitable intermediate equipment for trapezoidal ditches	Appropriate using graders for V-shaped ditches, and using back-hoes for trapezoidal ditches
Maintenance of culverts	Appropriate using labour gangs, but transport may be needed	Inappropriate	Inappropriate
Vegetation control	Appropriate using lengthmen or labour gangs (transport may be needed for gangs)	Tractor and mechanical mower	Appropriate using mechanical mowers for grass cutting on shoulders
Renewal of road furniture	Manufacture and placing of furniture is possible, but requires transport of furniture to site	Transport[1] to support hand manufacture and placing of furniture	Manufacture, transport and placing of furniture is possible

Note:
(1) Transport and haulage by tractor and trailer are unlikely to be economic for distances greater than about 10 km.

Table 8.4 *Roads 2000 comparison of partial rehabilitation methods*

Method	Direct costs (US$/km)	District overhead (US$/km)	Total cost (US$/km)
Labour-intensive	1447	350	1797
Intermediate	1188	430	1618
Equipment-based	1555	430	1985

Source: Kenya Roads 2000 programme.

8.4.4 Availability of labour and equipment

Labour-intensive methods are only appropriate where an abundant supply of labour is available. The ILO suggest that labour supply is only likely to be a problem where population density is less than 5 persons/km^2 (Vaidya, 1983). In addition to this demographic factor, labour availability is also affected by the availability of alternative employment; seasonal and climatic factors have an impact in some areas which might result in available labour being scarce: harvest times being an obvious example. There are also socio-cultural factors which result in labour not being available or willing to work on roads, even though it is relatively plentiful. Decisions must be taken on a area-by-area basis, recognising that the technology appropriate may also vary depending on the time of year.

Many developing countries and those with emerging economies are faced with major problems over the procurement of spare parts and the maintenance of mechanical equipment. This results in a low level of equipment availability that restricts the amount of road maintenance that can be carried out. Box 8.6 provides a simple example showing the consequences of wasting equipment resource costs. Thus, given the intractable difficulties of keeping mechanical equipment operational in many countries, there is a strong case for using labour-dominated methods (labour-

---

Box 8.6   *Substitution of equipment by labour in developing and emerging countries*

Typical ownership costs for road maintenance equipment are US$50/hour. This contrasts with casual labour costs that are often less than US$4/day in developing and emerging countries. Thus, for every piece of maintenance equipment that is unavailable for one day because of lack of spare parts, or other reasons, the equipment ownership costs would pay for one day's employment of approximately 100 casual labourers.

If a typical roads administration has 500 pieces of maintenance equipment that are awaiting repair and unavailable (in many of the countries in this category, the actual figure is considerably greater than this), then the daily ownership costs of this equipment is of the order of US$200 000. Assuming labour costs of US$4/day, then this sum would pay for the employment of 50 000 casual labourers. Assuming that one 'lengthman' is required for every half kilometre of road, this sum would pay for the maintenance of approximately 25 000 km of road.

intensive methods supported, where appropriate, by intermediate technology equipment) rather than more traditional mechanical equipment.

8.4.5 Domestic resources

Intermediate technology equipment that is manufactured locally should be easier to repair and keep operational than imported equipment. Several studies and programmes have demonstrated that tractor-based technology is likely to deliver higher equipment availability than the use of traditional equipment-intensive road maintenance. Not only is this approach more sustainable than the use of heavy equipment, but it is also less demanding on the use of imported resources, thus conserving foreign exchange.

8.4.6 Discussion on choice of technology

The above discussion makes a compelling case for the use of labour-dominated methods for carrying out road maintenance in those countries or regions where labour costs are below the break-even threshold. However, such a conclusion does depend on the individual maintenance activity, and on other factors. Table 8.5 summarises these discussions. It considers separately the recommended technology in those areas where labour is abundant and available to work, and in those where labour may not be available for one reason or another. Notes appended to the table indicate variations to the basic recommendations.

8.5 Equipment management

8.5.1 Basic principles

Several works activities rely on mechanisation to obtain levels of efficiency that are necessary for effective operations. The costs of owning and operating equipment are high, and often represent the largest single component of the cost of the works. Lack of working equipment has been cited as one of the most significant factors in road maintenance operations that are inefficient (World Bank, 1988). It is therefore crucial that equipment operations are well managed.

Equipment management is a specialised operation, and should therefore be undertaken by mechanical engineers. However, it is important that maintenance managers understand the basic principles involved if effective and efficient operations are to result from its use. One basic principle is that equipment should normally be managed as an autonomous unit, separated functionally from the maintenance or works unit. The equipment unit should be responsible for the procurement, repair and disposal of all vehicles and equipment, and should rent these to the works units according to commercial principles. Detailed guidance on the setting up and operating such an equipment unit has been given by Lantran and Lebussy (1991) of the World Bank, and this section provides a summary of their recommendations.

Table 8.5 *Recommendations for maintenance technology*

Maintenance activity	Maintenance technique	
	Situations where labour is generally available	Situations where labour generally unavailable
Patching of unpaved roads	Labour-intensive patching, with winning and haulage[1] of materials using intermediate technology methods for traffic levels less than about 50 veh/day; inappropriate for higher traffic levels	Inappropriate (use grading)
Grading of unpaved roads	Intermediate technology method using towed graders[2] for traffic levels greater than about 50 veh/day; not undertaken for lower traffic levels, where only patching should be used	
Regravelling of unpaved roads	Hand spreading of material for traffic levels less than about 50 veh/day; spreading by towed graders[2] for higher traffic levels although, for short lengths of road, the use of any spreading equipment may not be cost effective; haulage by intermediate technology methods[1]; spot regravelling is preferred to full regravelling	
Maintenance of side drains	Labour-intensive methods	Intermediate technology method using towed graders[2] for V-shaped ditches; and equipment-based methods using back-hoes for trapezoidal ditches
Maintenance of culverts	Labour-intensive methods, but transport to site may be needed[1]	
Vegetation control	Labour-intensive methods	Mechanical mowing of grass shoulders
Renewal of road furniture	Labour-intensive methods with intermediate technology transport[1]	

Notes:
(1) Tractor-based transport can be used for haul distances up to about 10 km, but truck-based transport is likely to be cost-effective for distances greater than this.
(2) Equipment trials are suggesting that towed graders can be substituted for motor graders for this activity, but motor graders can also be used.

8.5.2 *Autonomous management*

The functional separation of equipment management from works operations increases the specificity of the operation. In commercial organisations, the supply of equipment is also subject to competition, and there is no reason why a competitive element cannot also be introduced into public sector equipment management organisations (see Chapter 9). As discussed in Chapter 2, specificity and competition are the fundamental requirements for effective and efficient operations. In addition, equipment organisation needs to be set up in such a way to permit:

- The achievement of long-term corporate goals and planning mechanisms
- Tight management to minimise costs and maximise revenues
- The ability to respond quickly to market demands and to operate in a commercial manner.

The commercial management of equipment requires that rental rates are charged for the use of equipment that are sufficient to cover all of the operating costs of the equipment units, and to provide a fund from which replacement and additional equipment can be procured in the future. Such an approach tends to encourage optimum use of equipment because:

- The costs of operating equipment are explicit (and large), so there is pressure on works units to utilise only equipment that is actually needed at any time, and to obtain the maximum productivity from equipment on rent, otherwise unnecessary expenditure is incurred.
- No rental is received by the equipment unit for equipment not being used by works units, so there is an incentive to repair broken down equipment quickly and get it back into operation, and also to dispose of old and unwanted equipment.
- The equipment unit can develop a long-term strategy for equipment supply, and can respond to demands from customers in a flexible manner.

Even for public sector organisations, the equipment unit needs to be independent of government procurement rules and should, for example, be able to operate its own bank account. This enables spare parts to be procured in an efficient manner. Their method of procurement has to recognise that speed of delivery is an important aspect when customers are kept waiting because of slow repair times, and rental revenue is being lost. The equipment unit needs to be able to negotiate and set up arrangements with suppliers that are economical and efficient. Similarly, the unit must have the ability to set rental rates that reflect the true cost of owning and renting equipment. It also requires the flexibility to adjust these to reflect changes in the financial environment, such as inflation and currency exchange rate fluctuations, if adequate funds are to be generated for equipment replacement. A method of equipment costing suitable for this was described in Chapter 4. Finally, the unit should also have the ability to dispose of uneconomic equipment easily, and to retain income generated from this.

8.5.3 *Organisation of maintenance and repair*

Equipment maintenance operations can be conveniently organised according to their technical complexity into a number of 'levels' that are suitable for undertaking at different locations. A typical example would be:

- *Level 1 operations*
  Maintenance carried out at the road works site, and at local service facilities if distances to work sites are short
  – control of lubricants, cooling fluids, pressures in tyres, turbochargers, etc., and visual inspection of machines.
- *Level 2 operations*
  Work sites and local service facilities, as above
  – replacement of oil filters, hoses, and accessories that do not require tuning or specialised tools (e.g. alternators, starters, water pumps), and minor repairs.
- *Level 3 operations*
  Fixed workshops, including service facilities or regional workshops with towing and maintenance equipment
  – replacement of major components, such as engines and transmissions, and of accessories requiring special tools.
- *Level 4 operations*
  Large fixed workshops (central or regional)
  – repair of major components not requiring major milling or manufacturing, such as replacement of pistons and cases, and tuning of injector pumps.
- *Level 5 operations*
  Central workshops with highly qualified mechanics, or authorised dealers
  – repair of major components, with the ability to undertake major refurbishment operations, such as straightening cylinder heads and reboring.

The actual size, number and role of the workshops will depend on the size of the road administration and the types of works for which it is responsible.

The size of the equipment fleet will depend on the current and possible future demand for rental equipment. It will also depend on local circumstances and transport conditions. The above structure can accommodate 125–250 machines and 250–450 vehicles. In some situations, it may be preferable to set up several regional equipment units instead of a centralised operation. However, because of the fixed costs associated with workshops and offices, there is a minimum size of operation that is viable from a business point of view. This minimum number is likely to vary from over 100 machines and vehicles in large areas where equipment utilisation rates are moderate, down to about 60 machines in highly competitive environments. There is also considerable scope for contracting out workshop activities, particularly for Level 4 and 5 operations.

8.5.4 Administrative and commercial management

The commercial ambitions of the equipment unit should be realistic, since customer demands can be highly volatile and very diverse. The best strategy is normally to base equipment provision on that for which there is strong and steady demand, rather than meet the needs of all potential customers in terms of either volume or type of equipment. The diversity of the fleet owned should be limited by the cost-effectiveness of its operation, with peak demands and special needs being met by hiring from other organisations wherever this is possible.

Rental rates

Setting rental rates is a key issue for an equipment unit, and the principles to be considered when doing this include:

- True costs should be known
- Revenues should cover costs
- Revenues should be easy to calculate
- Rental rates should encourage customers to return machines that they are not using; this will allow the unit to allocate these to other customers.

Rental rates can be based on the methods of operational costing described in Chapter 4.

Spare parts management

Spare parts management is also a crucial element of equipment unit operation. The policy should normally be to maintain an adequate stock of fast moving parts, and to arrange expeditious supply of other parts from the manufacturer or from dealers.

Cost control

Cost control is essential for the equipment unit because it will not be able to rely on subsidies from its parent road administration. Separating the equipment unit functionally from the road works organisation facilitates this, since costs are explicit and not mixed up with the costs of other operations. The equipment unit will need to have a system for collecting cost data, and for classifying these costs by the type of machine in order to determine appropriate rental rates. There are three main categories of cost: running cost, depreciation, and non-allocatable general expenses. Running costs are relatively easy to monitor and control, but the other categories are discussed further.

Financial depreciation involves collecting enough revenue during the life of a machine so that it can be replaced when needed. As such, it is recommended that the depreciation account is separated from the current management account, and that revenue is recorded in the currency used for the purchase of the machine. Exchange rate variations should be reflected. Yearly depreciation forecasts should be consistent with realistic life-spans and the replacement cost of new machines, and these should feed back into the rental rates. Realistic residual values should also be taken into account, and revenue from equipment disposal should be placed into the depreciation account. Funds accumulating in this account are then available for equipment replacement. Such an approach also makes it easier to keep track of costs and revenues, and to update rental rates.

It is not necessary to allocate all costs directly to particular machines. General expenses and labour costs should be collected overall, and allocated to machine types using simple formulae. In a rental rate calculation based on the operational costing method described in Chapter 4, the cost of mechanics and lubricants would be related to the replacement cost of an equivalent new machine.

8.6 Monitoring

8.6.1 Issues to be considered

As noted in Chapter 1, the purpose of monitoring is to provide feed-back to the management process so that, when the next cycle of management takes place, it can learn from past experience. Monitoring should be undertaken irrespective of the method of delivery of the works.

The elements of the management cycle should be linked together in a coherent and compatible series of steps (see Chapter 1). Monitoring should provide a check that this does in fact occur. Under conditions of financial constraint, the works programme will not reflect objectives automatically, so that objectives may require modification to enable shortfalls to be redressed. Once objectives have been set for the year, or other time period, the following questions should be answered (Local Authority Associations, 1989):

- Are the objectives and desired level of service being achieved?
- Are the works being completed satisfactorily?
- Is value for money being obtained?

The three key areas over which control needs to be exercised are:

- Quality of work
- Final cost of work
- Duration of work.

Where work is being undertaken under a fixed price or an admeasure form of contract (see Chapter 9), then cost control is automatically delegated to the contractor. However, variations and potential for claims should be minimised. An opportunity cost arises out of delays in that resources cannot always be deployed onto other productive work. Quality of work is also very important and can have a considerable influence on durability and, hence, costs to the road owner.

8.6.2 Key indicators of effectiveness

Important indicators of effectiveness include:

- Volume of the various works carried out against that planned for each activity
- Trends relating to changes in network condition
- Expenditures per kilometre on different road classes for different treatment types
- Value and proportion of maintenance carried out on a year-by-year basis
- Cumulative amount of network treated or upgraded.

Other performance indicators were listed in Chapter 2. Monitoring should be a regular and on-going activity for all supervisory staff in the roads administration. A formal review of effectiveness, under the above headings, should be undertaken annually.

8.6.3 Value for money

Two important indicators are:

- Out-turn unit costs of works undertaken
- Time and cost over-runs.

Unit costs for tendered works cannot always be compared directly, as contractors' unit price bids may vary according to methodology and pricing strategy (see Chapter 4). Scope of works will also affect unit rates. However, some comparison should reveal trends and possible need for further investigation.

Time and cost over-runs may arise from many possible causes. These are not necessarily confined to occurrences during implementation of the work. They may have resulted from lack of definition of the works or other problems that have developed as a result of the works preparation phase. Reasons for over-runs should be identified to determine whether or not it would be worthwhile to instigate improvements in any of the various procedures between that of preparation and start of work.

A review of unit costs for the various activities should be carried out at least annually to update the values used in the estimating procedures for the subsequent year. The monitoring process should be used as the basis for seeking a gradual improvement over time of all aspects of road management, from planning through programming and preparation to operations.

8.6.4 Approach to monitoring

Monitoring involves site visits and desk reviews.

Site visits are important because they enable the road manager to become thoroughly familiar with road conditions throughout the region and to provide a first-hand check on the methods used and on the quality obtained on the works carried out. The presence of the engineer on the spot means that advice can be given on problems as they arise, and regular visits to site should boost the morale of road gangs and improve their standard of work and their output.

Desk review is an office task that involves reviewing all the maintenance documentation, such as inspection reports, completed work-sheets, estimates of unit costs and the like. It provides the opportunity to assess the performance of the works programme and the effectiveness of the system of management. Desk review also involves benchmarking against previous year and other similar operations.

9 Procurement and Contracts

- Procurement options
  - competitive procurement
  - construction and periodic maintenance
  - routine maintenance and emergency works
  - client functions
- General procurement issues
  - procedures and forms of contract
  - qualification
  - bidding documents and contracts
- Contract management
  - management principles
  - works supervision
- Contractor development
  - preliminary study
  - development programme
- Contracts for in-house works
  - improvement of existing operations
  - competitive procurement of activities

9.1 Procurement options

9.1.1 Options available

Various options are available for undertaking maintenance works. These include execution by in-house organisations, by external contractors, or by a combination of the two approaches. Similar options exist for the management of the works. Where work is undertaken by external contract, the following options are available:

- Traditional contracts
- Contracts for a fixed amount of work or a term contract
- Management contracts
- Design and build.

The use of competition to increase effectiveness and efficiency was discussed in Chapter 2. This chapter now discusses competitive procurement of maintenance, and the ways that this can be achieved using either in-house staff or the private sector.

9.1.2 Competitive procurement

It was noted in Chapter 2 that in many government organisations it is difficult to provide work incentives, and this is a particular problem in many road maintenance departments. The result is that work is often carried out inefficiently, and the quality and quantity of output is lower than it could be. Transferring work from the public sector to the private sector is unlikely, in itself, to have a beneficial effect. The problems remain the same: the essential difference is one of management. But, as a broad generalisation, private companies have simpler objectives than public sector organisations. They have sharper and more immediate motivation, and are often able to operate with greater flexibility. These characteristics make them better suited to problem-solving (EDI and ECA, 1991).

One way of obtaining greater efficiency is to introduce competition into operations. This can be done either by making greater use of existing private contractors, or by allowing public sector agencies to compete with the private sector. Contractors may achieve greater efficiency and lower costs because of competitive pressures that are unlikely to be present in a government organisation. Competition for road works can be expected to (Lantran and Morse, 1995a):

- Secure the best selection of contractors with regard to price and quality of works (economic effectiveness)
- To adapt supply to actual and varying demand for works, as contractors enter and exit the market (economic efficiency).

Some World Bank experience of using contractors is given in Box 9.1.

Box 9.1 *Use of contractors*

The use of local contractors can increase efficiency and reduce costs if they can be mobilised to undertake road works. Market and competitive forces tend to act more strongly on private firms – especially the small ones – than on public sector agencies. Their objectives are usually simple survival and profit and are not clouded by political consideration, and their relatively small size can increase their flexibility. These factors motivate them strongly to use staff efficiently and maximise the use of labour rather than capital equipment to conserve resources.

Source: EDI and ECA (1991).

9.1.3 *Contracting out construction and periodic maintenance works*

In many countries, delivery of new construction works and periodic maintenance have traditionally been carried out by contract. Contractors have also been used for the supply and haulage of gravel and aggregate, as well as for carrying out minor construction and improvement works.

The use of contractors for undertaking such works is relatively straightforward, and there is considerable experience of the issues involved. However, it should be noted that procedures for work execution on trunk, urban and rural roads are different. Contract procedures, such as those produced by the *Fédération Internationale des Ingénieurs Conseils* (FIDIC), favour the larger and more complex works. These may need to be modified to suit smaller sized works, such as might be needed for some periodic maintenance activities. When procuring works in this way, it is important to ensure that a real market exists, and that prices are not controlled by cartels.

Novel forms of contract are increasingly being introduced, with the aim of enhancing effectiveness and efficiency. An example of the use of 'lane rental contracts' is described in Box 9.2.

Box 9.2   *Lane rental contracts*

Lane rental contracts were introduced in the United Kingdom by the Department of Transport (DOT) in 1984 for major projects to speed up maintenance works on the busier roads and to reduce delays to road users. The basic contract arrangement is for a bonus to be paid if the contractor finishes the works before the contract completion date, but a charge is imposed, at the same rate as the bonus, if the contractor is late. Bonuses paid on individual projects have ranged from £5000 (approx. US$8000) to £1 million (approx. US$1.6 million). Both bonus and charge are based upon an assessment of the economic costs to road users of delay caused by the works. Two variant systems, continuous site rental and lane-by-lane rental, were introduced one year after the first introduction. The basic types of lane rental contracts are:

- *Bonus/rental charge* – the contractor
  - tenders a price for the work and a time of completion
  - receives a bonus or pays a charge according to the number of days that the work is completed ahead of or after the contract period
- *Continuous site rental* – the contractor is charged a daily rental for each day that there is possession of the site
- *Lane-by-lane rental* – the contractor is charged according to the number of lanes occupied

The form of contract used is a modified version of the standard ICE contract, where liquidated damages are replaced by a daily charge for late completion of work related to the cost to road users.

In setting the appropriate level of bonuses and charges, the DOT considered that it would be too costly to pay bonuses at the full daily delay rate, and that a lower rate would still give adequate incentives. They also took into account the probability that charges based on the full daily rate might not be recoverable, or might deter companies from tendering. It was, therefore, decided that the level of bonus should normally be set to cover 50 per cent of the daily delay cost plus 100 per cent of the daily site supervision cost. Later, the DOT limited the daily bonus and charge rates to a minimum of £2000 and a maximum of £25 000 (approx. in the range of US$3000–US$40 000).

Between 1984 and 1989, the DOT let about 100 lane rental contracts worth £250 million (approx. US$400 million). It is estimated that these contracts saved more than 2400 days of lane closures, representing an economic saving of some £50 million (approx. US$80 million), at an additional cost of £8 million (approx. US$13 million) paid in bonuses. Although there is some debate about the actual size of the benefits, detailed comparison of contracts in 1987–88 estimated that the average rate of spend on lane rental contracts per week was 81 per cent higher than that for a conventional contract; confirming that lane rental contracts have speeded up work substantially and reduced delays to traffic.

Adapted from: National Audit Office (1991). Crown copyright 1991.
                    Reproduced by permission of the Controller of H. M. Stationery Office.

9.1.4 Contracting out routine maintenance and emergency works

The use of contractors for routine cyclic and reactive maintenance poses more problems: works are often difficult to define and measure, and work may be spread over a wide area, making supervision and monitoring difficult. Contractual arrangements have been introduced in some countries to deal with this situation in an attempt to obtain greater effectiveness and efficiency. Routine works are procured on a competitive basis, typically using 'functional' rather than 'procedural' contracts, and often appointing consultants or contractors to manage a defined part of the network on a fixed term basis. Successful schemes of this nature have usually involved close collaboration between road administrations and contractors in defining the work to be done and in planning the work programme (Robinson, 1997).

Where such contracts have been introduced, cost savings are often quoted as being in the range 10–15 per cent (Harral *et al.*, 1986; Madelin, 1994b). As an example, some experience from Africa, quoted in Box 9.3, claims an even higher saving, although some of this is attributable to using an appropriate choice of technology. A particular type of maintenance contract that utilises 'lengthmen' has been used in some countries, and some experience of this is summarised in Box 9.4.

Box 9.3 *Cost reductions by using contractors in Ghana*

The cost of rehabilitation works by contract using intermediate technology was running about 20 per cent cheaper in financial terms than the equivalent capital-intensive work being done for DFR (Department of Feeder Roads). In addition, the foreign exchange component is now reduced by about 50 per cent from around 60 per cent to around 30 per cent.

Source: EDI and ECA (1991).

Box 9.4 *Lengthman contracts*

Principles of the approach
The use of individual contractors or lengthmen to undertake routine maintenance over specified lengths of road has been a recognised management technique for many years. Whereas this approach can be used for routine works, periodic works still need to be carried out by special units or contracting companies. Emergency works are carried out under various contractual arrangements when the scope is beyond the capabilities of the lengthman contractor.

An example of this approach is in Kenya, where ex-construction workers were chosen as lengthmen on a contract basis for each section of road, typically 1.5–2.0 km in length. Lengthmen were provided with hand tools and supervised once a month by an overseer to monitor condition of the road and to authorise payments for satisfactory work. The payment was based on the contractor carrying out 12 days of work per month on days of their choice. The contractor could be replaced if performance was consistently below requirements. The system enables maintenance to be achieved throughout the year on each section of road.

(continued)

Box 9.4 (*contd.*)

Thus, responsibility for the maintenance of each section of road lay completely with one person who required minimal logistical support. The contractors lived adjacent to the road and, therefore, did not require government accommodation or transport, which could consume considerable resources in a traditional equipment-based system. An attraction of the system was the comparatively low level of equipment required which considerably lessened support problems. This was coupled with a low foreign exchange component which was estimated to be only 10 per cent. This compared with a typical foreign exchange requirement for equipment-based routine maintenance of 50 per cent. The lengthman system also created productive paid employment in rural areas where there were few employment opportunities. Living at home with the family, and the part-time terms, also provided the opportunity for the contractor to work on their own land as well.

Experience of operation

It was found that the maintenance system did not attract the same level of interest as had the construction works. It was erroneously assumed that the local administration and people would bring pressure to bear on the contractors to maintain the roads to a good standard. The contractor's appreciation of the maintenance requirements was taken for granted. The need for training and supervision was underestimated, and mechanical problems with the supervision vehicles had an adverse effect.

Although the system proved to be effective, the need was recognised for good direction and control of the contractors:

- In particular, there is a need to determine the maintenance requirements under various conditions of rainfall, alignment, pavement/soil type, and traffic
- Methods of determining required maintenance resources and their deployment, direction and control need to be developed
- Arrangements for dealing with urgent works, such as wash-outs and culvert breakages need to be formulated
- The methods of identifying spot regravelling and full regravelling requirements need to be determined, and the various options for carrying out his work need to be investigated
- There is a requirement to ease the supervision burden of the maintenance overseers because of the minimal time that they can allocate to each contractor and the mechanical problems that will always exist to a degree, even with the low equipment component of the system
- Methods of training, directing and monitoring the headmen, responsible for a small number of contractors, need to be considered carefully
- On the technical side, there is a problem of maintaining a satisfactory longitudinal profile, especially for the wider, more heavily trafficked minor roads
- There is a question of the safety of lengthmen working on the carriageway of more heavily trafficked roads
- Consideration needs to be given to the use of simple tractor-drawn mechanical graders or drags for maintaining the running surface in these circumstances; with pot-hole patching support and all off-carriageway work by lengthmen.

Adapted from: Jones and Petts (1991).

In principle, there is no reason why emergency works cannot be undertaken by contract, and this is done in several countries. Either call-off contracts can be set up for the purpose, or all maintenance contracts can include clauses obliging the contractor to undertake emergency works, when directed, at day-work rates. Provided that enhanced rates are used for emergency works, and that penalties are provided for failure to respond within given time frames, then this approach can be effective. With such contracts, the client must always retain the ability to decide on use. However, some road administrations prefer to undertake emergency works in-house because they consider that they themselves are in the best position to provide the immediacy and flexibility of response in such situations, and to change priorities at short notice. It should be recognised that the cost of keeping such an organisation in-house is relatively high.

9.1.5 Contracting out the client functions

Scope of activities to be considered

It is also common practice for road construction and periodic maintenance works to be supervised by the private sector. Standard supervision contracts, such as those published by FIDIC, envisage the use of a supervising engineer who is independent of both the client and contractor. Such an approach can have all of the benefits of using the private sector under competitive arrangements, and can help to ensure a professional and unbiased approach.

Maintenance management has been put out to contract in some countries, and cost savings have been reported where consultants and contractors have been used. Consultants and contractors can also be used for tasks such as network referencing, inventory collection, traffic counts, axle load surveys, and condition surveys carried out both by manual methods and by machine. These works are relatively easy to define and, in several countries, specialist contractors have evolved for these activities who offer significant savings in cost. Similarly, for bridge management, consultants and contractors can often bring specialist skills and equipment to the tasks of bridge inspection and assessment. Indeed, some consultants specialise in the development and implementation of road management systems, and have skills in this area that are often not present within public sector organisations, particularly smaller ones.

Rather than try and develop public sector skills in all of these areas, it will often prove more cost-effective to use specialist consultants and contractors. This removes the need to develop and retain specialist skills within the client organisation, particularly where there is only a limited demand for these. Use of the private sector also gives access to state-of-the-art technology without the need for investing directly in development costs. Such an approach reduces the need for institutional development within the public sector.

Materials testing can also be carried out by specialist laboratories operated by the private sector. Tests can be carried out on a commercial basis, and competitive pressures between laboratories help to ensure that prices are controlled. With such an approach, there will normally be the need for a national accreditation process to ensure that laboratories have the skills and equipment to undertake certain tests to a satisfactory standard.

Moving over to competitive procurement makes the need for standards to be more explicit. However, making these too precise can reduce flexibility and be counterproductive.

Available models for contracting out the client function

Four models can be considered for contracting out this function:

- Agency agreements
- AGETIP
- Management of selected roads
- Management of the entire network.

Agency agreements

An agency is an organisation with delegated authority from the road owner for management of part of the network. Traditionally, agencies have been local authorities who manage part of the main road network on behalf of the national road administration. Agencies have, in the past, operated under a framework agreement that sets out activities to be performed, but with considerable discretion to undertake work as they consider appropriate. Typically, activities have included designing, supervising, and managing maintenance, improvement works and winter maintenance, and ensuring that the network is maintained to national standards. More recently, agency agreements have moved further towards performance-based agreements, rather than simply specifying activities to be undertaken.

AGETIP

This is a specific kind of agency arrangement used extensively in francophone Africa (Lantran, 1990–1993). The AGETIP is a contract executing agency set up to execute donor-financed infrastructure projects. The agency generally has a board of administration composed of well-known figures, but not including government representatives. A general manager is appointed by the board, and line managers and other staff are hired under private sector terms and conditions of service. The agency is set up as a private, non-profit association and pays no taxes. It works on behalf of local authorities who delegate certain functions to the agency. The local government usually reserves the right to select the projects, and the agency then:

- Recruits consultants to carry out detailed engineering
- Invites bids and awards contracts for supervision and works
- Manages the contracts
- Pays the contractors directly from a special account opened in its own name.

The agency is subject to bi-monthly management and financial audits, and an annual technical audit.

The overhead cost of the AGETIP in Senegal, for example, has worked out to be about 5 per cent on a turnover of US$55 million, excluding fees paid to consultants for designing and supervising the 330 projects. The advantages of the AGETIP are that it:

- Gets around cumbersome government procurement regulations
- Pays high salaries, and therefore attracts well-motivated, high-quality staff.

The disadvantages are that:

- The arrangement is not subject to competitive bidding (note that consultants, managing the district road network in Zambia under other arrangements, are only charging a fee of 2.5 per cent of the cost of the works, and this includes preparing and supervising the road maintenance programme).
- It is almost entirely dependent on donor funding.
- It probably hampers development of the local consulting industry by creaming off staff and monopolising all contract execution work for itself under a tax-free operating environment.

Management of selected roads

This method is being used in a number of countries, including Argentina, Australia, New Zealand and the United Kingdom. In the United Kingdom, the process started in 1986 when the Department of Transport decided to package parts of the motorway network into commissions, and then to invite bids from consultants to take on the responsibility for maintaining all roads and related structures within the commission to a prescribed standard. The winning consultant then organises a competitive term contract between the owner and the contractor, who then carries out all work on instruction from the consultant. The adoption of this approach to maintenance management has led to a claimed 15 per cent reduction in management costs (Boddy, 1988). Most members of the existing maintenance work force transferred to the consultant's staff upon the award of contracts, so the approach provided competition and increased levels of specificity without significant job losses. The Australian experience has been described by Smith, R. B. *et al.* (1994).

Different organisational models can be used for management under these arrangements, as shown in Figure 9.1. In the United Kingdom, this method is now being used to embrace the traditional road agencies, and this is described in Box 9.5.

Management of the entire road network

The final model involves contracting out the management function for the whole road network under the jurisdiction of a selected road administration. In industrialised countries, this is generally being done to increase efficiency, and as part of the redefinition of the role of government. In developing and transitional countries, it is mainly being done to ensure that small urban and district roads are managed by a competent body which remains answerable to the local district council. This model is being used by some small municipalities in the United States, and at the county council and district council level in the United Kingdom, where it is called 'externalisation'. It is also being used for both urban and rural district councils in Zambia. These arrangements offer considerable potential for dealing with small road networks.

256 *Road Maintenance Management*

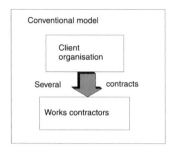

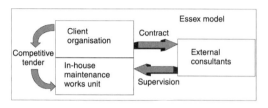

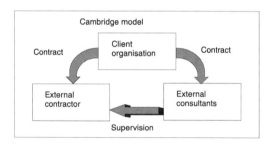

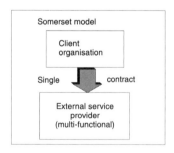

Figure 9.1 *Models for contracting out road maintenance* (*Source*: D. Robinson of W. S. Atkins Consultants)

9.1.6 *Considerations for maintenance by contract*

Advantages and disadvantages

The construction and supervision of capital works by contract is a well-established practice in many countries and is not discussed further. The remainder of the chapter

Box 9.5 *Example of contracting out the management of selected roads*

In April 1994, the Highways Agency was formed in the United Kingdom by hiving off part of the Department of Transport as an executive agency. It was tasked with 'securing the delivery of an efficient, reliable, safe and environmentally acceptable trunk road network'. One of its first tasks was to review the existing agency arrangements, keeping in mind two important objectives:

- To establish a more customer-focused approach to the delivery of service
- To make maximum use of the existing network (i.e. to look at the strategic purpose of trunk roads and to develop their use to maximise returns)

The Agency had the direct responsibility for the maintenance and improvement of the trunk road and motorway network through 91 agencies, of which six were consultants. Differences between the agency arrangements of local authorities and those of consultants are shown below.

	Local authority	*Consultant*
Statutory functions	Delegated	Non-delegated
Agency Agreement	On-going	Contractual
	Non-competitive	Competitive
Payment	In advance, based on estimates	In arrears, based on actual costs

In 1996, changes in agency arrangements were introduced for a number of reasons:

- Government policy decreed the need for more competition and the *Local Government Review* was expected to lead to changes in the structure of local government and to boundaries of local authorities
- New design, build, finance and operate (DBFO) projects were being taken out of the locally-managed network, possibly creating non-viable agencies
- Externalisation of local government engineering design groups and the imminent arrival of compulsory competitive tendering (CCT) meant that an extra and unnecessary layer of management was being created

In addition, the Highways Agency had been aware for some time that managing the motorway and trunk road network on the basis of administrative boundaries was not always the most appropriate way to meet strategic needs. As a result, the opportunity was taken to introduce changes. The principal change was that agreements were introduced for a fixed term of up to five years, and would be open to competition from both the public and private sectors. This involved replacing 91 agencies by 30 new ones based on strategic considerations, and of an appropriate size for medium-sized companies and local authorities to compete on equal terms. The larger size of agency allowed routes to be managed across local authority boundaries. They were also planned to provide a balance between competition and strategic management of the network.

Bids were accepted from:

- Local authorities
- Consultants
- Associations between local authorities and consultants, with either party as the lead or, effectively, as a joint venture

(*continued*)

Box 9.5 (*contd.*)

Legislation was changed to enable local authorities to compete and to allow private sector companies to have statutory powers. Contractual relationships are as shown below.

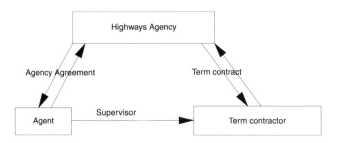

Performance incentives include potential extension of the agency agreement and the letting of smaller works directly to the contractor. The main performance criterion is the delivery of defined value for money. However, there was a need to address concerns in terms of:

- Would road users know who to complain to?
- The need for co-ordination of activities across local authority boundaries, and across the trunk and local road networks
- Potential loss of economies of scale.

Adapted from: Nutt *et al.* (1996).

---

focuses on the execution and management of road maintenance by contract. Some advantages and disadvantages of using contracts for these works are given in Box 9.6.

It is not always possible to obtain benefits from using contractors. Moving to the use of the private sector can reduce the flexibility to make changes through the year, or to vire funds, perhaps between overlays and winter maintenance. Madelin (1994b) notes that, when comparing a private contractor with an in-house organisation that is already operating efficiently, no evidence of savings is found. Unfortunately, the efficiency of the routine maintenance operations of many of the world's in-house organisations leaves much to be desired. In public sector organisations, savings can be made with separation of the client and contractor roles, since separation gives a clarity of purpose and a focus for management. Further savings can be made by revitalising the organisation of the in-house operation. The use of contracts for works carried out by in-house units is discussed in section 9.5.

Comparing costs

There are also significant difficulties in comparing the costs of in-house works with those carried out by contract. Costings used by in-house units normally exclude significant elements of cost, whereas contractors must employ full cost accounting, including elements for profit, equipment depreciation and down-time, supervisory staff expenses, office overheads and insurance. This was discussed in Chapter 4. For example, a cost comparison between an in-house operation and contractors was made in Kenya, and the results are shown in Table 9.1. In order to achieve this comparison on an equitable basis, the in-house figures needed to be adjusted to take account of the following:

- Inclusion of all vehicle and equipment depreciation and capital investment costs.
- Comparisons had to be undertaken
  - for an equal length of road works
  - for roads of equal standard, e.g. adjustments must be made for differences between cross-sections on the different roads evaluated.
- Materials acquisition costs needed to be included in the total sum in both cases.
- Contract management and training expenditures needed to be included in the total costs.

This particular comparison suggested that a saving of about 10 per cent was obtained for external contract works over in-house works.

Box 9.6 *Advantages and disadvantages of private sector participation in road management*

Advantages

- Strong incentives for improvements in performance and economy
- A more flexible operating environment in terms of managing resources, including greater flexibility in scaling resources to suit changing demands, thus facilitating improvements in cost-effectiveness
- Relief to the government from the burden of direct management responsibilities for large work forces and equipment fleets
- For maintenance works, the need to commit funds for maintenance contracts, with less likely diversion of resources to other activities
- Political support for adequate and more stable levels of funding for road works, provided the contractors can organise themselves into a reasonable lobby
- A better prospect for developing a lasting institutional capacity, in the form of a pool of local contractors skilled in providing efficient and effective services

Disadvantages

- Contracting may not decrease costs where redundant government establishment and work forces cannot be reduced or relocated to the private sector
- Contracting could increase costs because the very process of contracting and contract administration may require additional government resources, e.g. for measuring and certifying work quantities for payment
- Contracting may increase costs to the government where there is lack of effective competition in the procurement process, including abuses such as price fixing and corruption in contractor selection
- Government may not have the capabilities necessary to manage contracts properly
- Domestic contractors may not have sufficient capabilities, in terms of management abilities, technical skills, equipment, working capital, and other resources, that are necessary to ensure effective execution of maintenance activities
- Contractors may not be well placed to respond quickly during times of emergency, or to address small scale maintenance needs in remote areas

Adapted from: Harral *et al.* (1986).

Table 9.1 *Costs of improvement works by in-house works and contract in Kenya*

Improvement works	Cost/km (KSh)
In-house works	1 006 000
Contract	910 000

Conditions for success

The success of contracting also depends, to a large extent, on the institutional and macro-economic framework. Some institutional requirements are given in Box 9.7. Economic distortions, artificial exchange rates and inappropriate allocation of resources due to excessive centralisation have a negative impact on all producers, and particularly on contractors working for the government. There is a need to find the best methods for specifying targets and incentives that encourage achievement of what is needed, rather than others that maximise profits. There is also a need for adequate guidelines and procedures, since those developed for in-house works may no longer be appropriate. Other mechanisms for successful use of contractors are listed in Box 9.8.

In those countries where there is a weak private sector contracting capability, privatisation may also be an option for increasing competition. However, creating a separate enterprise that still enjoys a monopoly position is unlikely to provide the full

---

Box 9.7 *Institutional requirements for success*

Enhancement of the policy environment:

- Policies and practices for procurement and contract management
- Licensing and pre-qualification procedures
- Procurement regulations and practices
- Conditions of contract
- Contract management, including payment procedures and practices

Improvement of the construction industry business environment:

- Access to credit, guarantees, import licences, and foreign exchange
- Availability of equipment and spare parts
- Competition with publicly-owned firms

Development of institutions:

- Promoting agency
- Contractors' and consultants' associations

Promotion of research and development:

- Study on the macro-economic impact of the industry
- Performance criteria
- Laboratories

Adapted from: Lantran (1990–1993).

advantages, although effectiveness and efficiency are still likely to be increased because of a higher degree of specificity. In countries where the private sector is poorly developed, maintenance contracts can be useful in encouraging the local contracting industry and in providing an introduction to work in the roads sub-sector. However, it is important to ensure that contractors appointed do in fact have the capability actually to carry out the activities for which they have bid.

Box 9.8   *Practices and tools that help to achieve competition*

Stable financing over time
This can lead to a steady supply of work, and convince prospective contractors that there are good long-term business opportunities in the road sub-sector, and that this is therefore a market in which it is worth investing.

Informed contractors
Work programmes should be published well in advance of any call for bids, so that contractors can assess the attractiveness of the market where they are not present.

Registration and pre-qualification
Adequate pre-qualification procedures should be used for large contracts and registers used for small and medium-sized contracts. Registers should be maintained through annual revisions, and should take into account incidents such as accident record, poor performance, very late completion of works, solvency, etc. Newcomers should be able to register provided they meet published requirements.

Monitoring and review of competition
The list of potential and actual bidders should be reviewed periodically to assist in establishing a strategy for competition when launching programmes of work. The review should also take into account the actual levels of competition achieved.

Division of work into lots
Large lots will attract large firms; small lots will generally attract small firms. Slicing and packaging works into lots of an appropriate size and number can be a proactive way to encourage competition. When funds are tight and work programmes small, firms can be kept alive by splitting programmes into relatively small lots. When the good years return, larger contracts can be re-introduced.

Specialised lots
Splitting procurement along technical lines, such as for earthmoving, pavement works, bridges, etc, will attract specialised firms, whereas more general works will attract the converse. A particular case to consider is where one of the potential bidders owns the only quarry acceptable for the works. If they are allowed to bid, then the competition will be flawed if the winning bidder must buy the material from a competitor. In such a case, it is better to place a contract for the supply of material to the quarry owner, and then call for competition on the works.

Equipment
Equipment purchase can be a major problem for contractors, because of the financial resources required. There is a risk that equipment procured cannot be depreciated fully over an adequate period. Equipment leasing or renting can ease this burden.

Sub-contracting
It is better that contractors with limited skills or resources should be a sub-contractor rather than a full sized partner in a large contract. This arrangement assists in providing a level of

(continued)

Box 9.8 (*contd.*)

risk that can be managed. Where specialist contractors are used, there is no benefit in involving them in the management of other technical works. In cases where a specialist contractor has been selected by an owner, perhaps through selective competitive bidding, special arrangements need to be set up with the general contractor for works. Standard contract clauses should be proposed and publicised by clients in these situations.

Technical risk
Bidding documents and draft contracts for works should share risk equitably between client and contractor. Deliberate or inadvertent placing of an excessive burden of risk on to contractors can deter good contractors from bidding, or can increase prices.

Contractor financing
Measures that reduce contractors' financial burdens include monthly payments, advance payments, quick payment procedures, payments in foreign currency if appropriate, and fair escalation formulae. Reducing the financial burden on contractors enables more firms to bid and increases competition.

Multi-year contracts
It is sometimes appropriate for small contracts to be packaged together into multi-year contracts. This enables contractors to depreciate investments in the site and equipment over a longer period, and can increase competition in certain circumstances.

Cancelling bids
If few bids appear, or if prices are high compared to the engineer's estimate, clients can cancel bids. If a new bidding process is then started, the framework for competition should be changed, either by slicing or packaging the works differently, or by modifying specific features of the bid that have deterred competitors; this may include items such as financial risks, or completion time, for example. Sometimes it is not possible for works to be delayed, and the client is forced to accept the high price. In such circumstances, the price should be made public to attract other contractors to the potential profits, and to bid next time.

Adapted from: Lantran and Morse (1995a).

Where there is no tradition for undertaking road works by contract, operations of contractors have often been hampered by under-capitalisation. This is partly because of the difficulty in securing adequate finance since contractors are seen as a bad risk by the banking sector. The problem is compounded because of delays in payment for works executed and a lack of a consistent policy for contracting that makes it difficult to make long-term investments in equipment. A lack of adequate equipment support is an inevitable consequence of this. In addition, there is sometimes a lack of contract management skills in both the client and contracting organisations, and there may be a lack of technical skills. In some countries, adequate rules and regulations for contracts do not exist. For the efficient and effective management of roads by the private sector, the following conditions are required:

- Steady funding must be ensured by
  - political and economic stability to establish a climate of confidence and co-operation among operatives
  - provision of a predictable workload
  - timely payment for works carried out.

- Adaptable bidding procedures and contract documents to suit the nature of works to be undertaken.
- Accountability and transparency of bidding.
- Application of incentives and sanctions as applicable.
- Continuous monitoring and evaluation.

9.2 General procurement issues

9.2.1 Basic principles

Procurement should aim to achieve effectiveness and efficiency (see Chapter 2) which, in the case of contracting, means seeking to obtain the lowest price for goods or services that meets the purchaser's requirements in a manner that is rapid and routine for both buyers and sellers (Lantran and Morse, 1995b). Although equal opportunity and fairness are not direct objectives of the procurement process, they are prerequisites to achieving long-term effectiveness and efficiency. Equal opportunity ensures that all potential suppliers learn of bidding opportunities and are permitted to participate. Fairness tells bidders that their offers will be treated equitably, and that the award decision will be based only on their tender offer with respect to both technical content and price.

Open access is obtained by publicity in accordance with local conditions and traditions. This may include an advertisement of the potential bidding opportunity in trade journals, local newspapers or in an official gazette, advising contractor associations, posting on bulletin boards and the like. A consistent, reliable and readily available method of notification should be established.

The procedure for submitting bids, the methodology for evaluation and the procedure for contract awards should be known and followed consistently. It is customary for government procurement to be based on legislation that sets out the principal approach to be followed. Both official and private purchasers should have standard bidding documents and written procedures that specify how the procurement should be conducted. Consistency and predictability are important confidence building factors. Bidding documents that have been in use over a long period accumulate interpretations resulting from arbitration and court decisions that establish the ground rules for their application. If procurement is to be credible, some mechanism must be available to allow unsuccessful bidders to appeal if they consider that an error has occurred in the tendering process or evaluation. This could be an independent board established specifically to consider such appeals. In addition, every bidder should have access to the courts when other less formal dispute resolution procedures have proved to be unsatisfactory.

9.2.2 Procedures and forms of contract

Contractual arrangements

Payment methods used could have a major bearing on the cost of the works. It is therefore appropriate to consider this choice at an early stage. The following types of payment method are most common:

- *Lump sum*
  Payment based on a single price for the total work, or on the achievement of completed activities, groups of activities, or milestones.
- *Admeasure*
  Payment for quantities of completed work, valued at tendered rates in a bill of quantities or schedule of rates.
- *Cost-reimbursable*
  Payment for actual cost incurred by the contractor (requires 'open book' accounting) plus fee for overheads and profit.
- *Target cost*
  Payment based on actual cost, plus fee, plus incentive.

Choice of one or other type will be largely dictated by the extent of risk, the risk-carrying capability of the parties, and the extent to which design is complete at the tender stage.

Price-based contracts

In lump sum and admeasure contracts, the contractor carries significant risk and tenders have to be priced accordingly. When the contractor's risks are high, this results in extremely high prices or in some contractors being reluctant to tender at all. It can also result in unrealistically low bids where risk has not been properly allowed for by inexperienced contractors.

Cost-based contracts

In cost-reimbursable and target cost contracts, the client will bear most of the risk. Target cost contracts introduce an incentive for the contractor to work efficiently, aligning their objectives with that of the client to achieve the work cost-effectively. This is covered by setting a target price, usually by competitive tender, and agreeing a mechanism for sharing differences between the final actual cost and the final target price. Thus, if the work is completed at less than the target cost, a proportion of the cost saving is paid to the contractor as a bonus. But if the target cost is exceeded, at least some of the extra cost may need to be borne by the contractor. Where the contract is managed with 'open book' accounting, this provides the opportunity for client, consultant and contractor to discuss modifications to the works when these may be desirable. Such flexibility is desirable under uncertain conditions. This approach requires a measure of trust between client and contractor, and this has often been lacking in the past. Such methods of contract are also more demanding of senior site staff for both consultant and contractor.

Types of specification

There are two basic types of specifications that can be used in contracts:

- Procedural (or method) specification, where the client defines details of the work to be carried out (sometimes known colloquially as a 'cook-book' specification).

- Functional (or end-product) specification, where the client defines the result to be achieved by the work in terms of a functional or performance requirement.

Procedural specifications have been used traditionally for roads works. These reflect the high degree of competence of road administrations, and are relatively easy to specify and to measure. However, they have high supervisory requirements and do little to encourage contractor innovation, since there is little permitted flexibility for changing work methods, designs, or materials. Particularly for road maintenance contracts, where the amount of supervision required can present a problem for client organisations, functional specifications offer certain advantages. Defining performance standards in functional terms, such as road surface friction values to be obtained, means that supervision requirements are minimised, since it is only necessary to test the end result. Contractors can then determine the most appropriate way to meet the performance requirement that maximises the use of their own particular skills, equipment and use of materials. This approach also encourages contractor innovation.

The main difficulty with functional specifications is the need to describe and define the functional requirements for all activities. This is simplified by splitting works up into small units. Examples where this can be done are for maintenance works for grass cutting, ditching and surface repairs, where it is relatively easy to define the work in terms of the function of the end product, but can lead to problems of co-ordination between activities. This also facilitates the use of small and/or specialist sub-contractors. They may also have more incentive to innovate than firms with more diverse interests because of the need to retain market advantage in their specialist area. The problems of co-ordination can be minimised by using a 'generic' or a management contractor to take responsibility for all maintenance works on a road network on a 'term' basis. The contractor will undertake some work directly, but will also arrange and manage the inputs of specialist sub-contractors. However, this can result in reduced competition at the end of the term when the work is re-tendered.

Scope of works

Contracting for specific items of work, such as resealing, overlay or reconstruction of a specific length of pavement, are widely used, and there is considerable experience of this. However, particularly for road maintenance works, there is often a need for contracts to cover a wider scope of work. For example, Algeria, Belgium, Chile, Malaysia and some other countries use standard contract documents that may be different for major and minor maintenance works (Miguel and Condron, 1991). Routine and periodic maintenance operations are usually contracted separately. This practice is mostly used in Chile, Kenya and Pakistan, and is applied frequently in other countries to more complex periodic activities, such as pavement or bridge repair work. In Algeria and Brazil, maintenance contracts for specific road sections (on average about 250 km in Brazil) combine the execution of routine and periodic maintenance.

Some countries, including Canada (British Columbia), Malaysia, Sweden and the United Kingdom, have experience of including all maintenance activities on specific routes, or within entire geographic areas, in comprehensive maintenance contracts combining both periodic and routine works. Contractors additionally are responsible for managing the maintenance and operations programmes, including performing routine patrols and detailed inspections to identify needs, setting priorities, scheduling the work, and public relations. The contracts used in British Columbia and the United Kingdom are typically of between 3 and 5 years, whereas Malaysia uses contracts of 2 year duration. Contractors in these countries have stated that they consider a contract period of 5 years provides sufficient incentive to invest in costly and specialised equipment.

Incentives to contractors

Contracts can also be set up to include benefits to the contractor to reward good performance and, conversely, penalties in the case of non-performance. Cost-based contracts enable this to be done most easily. Cost-reimbursable contracts give incentives to contractors because their risk is minimised, although this form of contract offers no inducements to minimise costs. Target cost contracts perhaps provide a better basis for incentives since both the contractor and client have an incentive to minimise costs and to maximise performance.

In price-based contracts, liquidated damages are often applied when work is completed late, and this gives an incentive to the contractor to complete works within the contract period. However, there are seldom incentives to finish work early. Lane rental contracts are an example where incentives for early completion exist using this form of contract, and these were described in Box 9.2.

Partnering

Fundamental changes are presently under way in the approach to procurement by contract. For the past 25 years, the industry has been plagued with claims, litigation and cost over-runs. For example, in the United Kingdom at least one-quarter of all money paid to contractors on tendered contracts is either a claim or an extra, and is the result of a non-competitive deal. It has been claimed that contractors' profits would improve by 30 per cent if they did not have to cover delayed claims payments (Bolton, 1997). Changes are now under way to promote less adversarial relationships on engineering projects. These involve the concept of 'partnering' and other attempts at making clients and engineering professionals work together to ensure that contracts are undertaken as efficiently and cost-effectively as possible. The aim is to develop, within a contractual relationship, an atmosphere of co-operation and a team-work approach to projects between clients and contractors, thus reducing conflicts.

This approach is embodied in the *New Engineering Contract* (NEC) (Institution of Civil Engineers, 1996a). This requires, for example, an undertaking that parties to the contract will act 'in a spirit of mutual trust and co-operation'. The contract is non-adversarial in nature, and has the aim of promoting good management and quality information for employer and contractor alike. It is solution-focused rather than being problem-orientated. Users of the NEC have found that there are significant advantages and cost savings through the increased openness engendered by the contract.

The NEC contains the following options:

- Priced contract with activity schedule or bill of quantities
- Target contract with activity schedule or bill of quantities
- Cost reimbursable contract
- Management contract.

The partnering approach has been applied to road maintenance management (Robinson, 1997). The use of area management contracts covering a wide range of maintenance activities has provided the opportunity to integrate the provision of maintenance services in ways that were not possible under previous arrangements in the public sector. However, it has been found that partnering is not an easy option, although its use has generally resulted in improved levels of service and better value for money.

9.2.3 Qualification

It is important for the employer to be sure that bidders have the ability to undertake any work that is awarded. This is done by stipulating criteria, in advance, that bidders must meet. Typical criteria might include (Lantran and Morse, 1995b) evidence that the contractor:

- Has undertaken similar works in the recent past.
- Has an annual turnover many times greater than the value of the works being tendered.
- Is not over-committed to other on-going or upcoming contracts.
- Has the necessary financial capacity, including access to credit, to meet cash-flow requirements during periods prior to progress payments being made.
- Owns or has access to the necessary equipment.
- Can assign experienced personnel to the works.

Pre-qualification is often undertaken since this saves time in subsequent bid evaluation. With this, potential bidders respond to a pre-qualification document that describes the works, provides an abbreviated specification and contract conditions, and requests information on whether the contractor's operation meets or exceeds specified threshold criteria. Only those firms who submit pre-qualification documents and meet the specified criteria are invited to bid. Pre-qualification should not be undertaken for small works where the effort involved is disproportionate to the size of the works. In such cases, it is preferable to set up a permanent register of contractors that records qualification for different categories and amounts of work. Pre-qualification removes questions about contractor capability from the evaluation of bids. It ensures that only qualified contractors can bid, and removes the risk that marginally qualified contractors may buy their way into a contract with unrealistically low bids. Pre-qualification is also useful to obtain an indication of the interest from potential bidders, and allows questions of clarification to be raised before bidding conditions have become fixed. It also saves bidders the expense of submitting a full bid that will clearly be unacceptable.

Post-qualification is essentially the same process, except that the evidence that bidders can meet specified thresholds is submitted with the bid itself. The qualifications of the lowest bidder are then scrutinised to ensure that they are qualified to undertake the contract. Post-qualification is normally used for simple works and for the purchase of standard supplies from established manufacturers or suppliers.

9.2.4 Bidding documents and contracts

The following general principles are applicable to the preparation of bidding documents and contracts to assist in meeting requirements of effectiveness, efficiency and fairness (Local Authority Associations, 1989):

- Avoid unnecessarily complex documentation.
- Standardise contract procedures as far as possible.
- Contracts should enable standards and specifications to be enforceable in a clear and unambiguous manner.
- The use of functional specification contracts should be considered to improve quality of work.
- Size and scope of contract should aim to make the most efficient use of available resources, and spread fixed costs in an optimal way.

When preparing bidding documents and contracts, there is also a need to consider issues of insurance and contract securities. For example, the contractor is normally required to carry insurance against third party liability, and against loss to the works, plant and materials. Insurance requirements should be specified in the contract documents, and evidence that the required cover has been obtained should normally be sought from the contractor by the employer. Project owners can also take out insurance to cover losses that they might incur. However, many large organisations and most governments deliberately, or inadvertently, self-insure. Self-insurance assumes that the cost of any insurance premiums payable would be greater than any losses that might be suffered. But self-insurance also requires that the financial resources are available to cover any losses. Project insurance is likely to be more economical and to provide better cover for large projects. This would cover client, contractor and any sub-contractors.

Contract securities that are used in conjunction with competitive bidding and works contracts include the following.

Bid bond

This normally consists of a certified cheque, bank draft, stand-by letter of credit, or bond from a bank that provides an assurance that the bidder will honour their bid if it is accepted. Bid bonds are typically in the range of 1 to 3 per cent of the contract value.

Performance bond

Provision of a performance security or bank guarantee is normally required to be provided to the employer by the contractor shortly after the bid is accepted. This may be required in an unconditional form that must be paid without the employer proving reasons for any demand, or can be conditional where written notice is needed from the contractor, adjudication authority, court or arbitration award before payment can be made. Typical values are 10 per cent of the contract value.

Advance payment guarantee

This is a guarantee for the amount of any advance payment that is to be made. It is usually an unconditional obligation.

Retention of a proportion of payments is widely used as a guarantee against meeting contractual obligations. Retention money is paid either on completion of the works, or at the conclusion of a subsequent 'maintenance period' during which the contractor is responsible for correcting any defects that appear.

The above discussion does not imply that contracts are tailor-made for each project. The aim should be to select the most appropriate type of contract, from among those listed earlier, that will give the greatest chance of project success. The general principles listed in Box 9.9 should be adhered to when awarding contracts.

Box 9.9 *General principles for awarding contracts*

- Non-commercial matters should not be considered when drawing up select lists; bias is to be avoided
- Contractors on the list should be those able to provide the least expensive tenders while competent to undertake work to the satisfaction of the client
- Subject to the previous principles, each contractor on the select list should be given the opportunity to tender from time to time
- The number of contractors invited to tender should be between four and six; sufficient numbers for competition, but not so excessive as to waste effort and increase the level of contractor tendering costs unnecessarily
- Tenders should be received at a time and place notified in the tender invitation, opened in the presence of authorised representatives, recorded and evaluated in an unbiased way
- The road administration should not be bound to accept the lowest tender, if there are good reasons to expect that a better result would be obtained by accepting a more expensive tender
- Details of tenders should be made public, but commercial confidences should not be disclosed unreasonably

Adapted from: Local Authority Associations (1989).

9.3 Contract management

9.3.1 Management principles

The role of the contract or project manager is to deliver the contract to time, to budget and to the agreed quality. As such, it is principally management skills that are required rather than those of engineering. The contract manager may often need to stand back from the technical aspects of the work, delegating these to others, in order to focus on meeting contractual requirements. The contract manager should act as a facilitator to assist team members in completing their work to meet contract requirements. Skills will be required in planning and co-ordination. But, above all, excellent communication skills are necessary, both oral and written: good communication is fundamental to a well-run contract. The contract manager, ideally, should have leadership qualities (Adair, 1983) and should command the respect of the client and colleagues alike by strength of personality. The way in which the contract manager communicates with the client can do much to foster good relations.

Mobilisation is a key stage of the contract. Plans and procedures put in place at this time will have a key influence on how the remainder of the contract is run, and on its eventual success. Work plans should be based on critical path analysis (Freeman-Bell and Balkwill, 1993) to detail all actions required, and to identify the schedules and resources required to carry out the actions. A plan would have been prepared as part of the tendering exercise, but this will now need to be updated to reflect known project circumstances. Any schedule must be reviewed by personnel involved in the works since it is essential that they have ownership of it. The employer will also be interested in this schedule to enable the progress of the contractor to be reviewed more carefully.

The contract needs to be controlled against the two key management tools: the project plan and the project budget. The project plan, or work programme, is the basic time management tool. The project budget provides the basis of managing costs. The budget should be broken down by each activity in the project plan in terms of the resources of labour, materials and equipment. Cash-flow forecasts should also be made and used in conjunction with the budget to monitor expenditure and cash-flow over time against output. Project management software is usually used to assist with the on-going control of contracts and, in particular, management against the project plan. Accounting software may be used to assist with budget management.

Changes to the scope of a contract occur invariably throughout the contract's life for a number of reasons. Consequences of any change can be totally out of proportion to the actual change made. It is necessary to ensure that the full implications of any change are assessed before it is made, so that the effects can be incorporated into the work plans and budgets for all aspects of the contract. It is impractical to prevent change. What is needed is a mechanism for coping with its effects. A change control system needs to include the following elements (Open University, 1987):

- A defined baseline from which any change is referenced.
- A means of submitting a proposal for change.
- A mechanism for evaluating the impacts of change elsewhere in the contract, including effects on both time and cost.
- An individual or body with the authority to approve or reject submissions.
- A method of recording the approved, or non-approved, change.
- A method of publishing the change.
- A means of monitoring the implementation of change.

Only the NEC, among standard contract forms, meets all of these criteria.

9.3.2 Works supervision

The principal objective of works supervision is to achieve good quality work within programme and to budget, in accordance with the specification, drawings and other contract documents available. The effect of work on safety and liaison with interested parties (public relations) may be of specific importance to the supervising authority in some cases. The way in which work is performed is entirely the responsibility of the contractor, unless it conflicts with the contract documents or legislation.

Resources

The client organisation should endeavour to provide personnel resources with adequate relevant experience and training, together with testing and other back-up facilities to enable a satisfactory level of quality control to be achieved. Factors which affect the desirable level of supervision include:

- Scale and type of work
- Duration
- Capability of contractor
- Complexity of work
- Type of contract and associated procedures
- Third party liaison and safety requirements
- Implications of defective work going unnoticed
- Degree of definition provided by contract documents
- Traditional arrangements, which can be well-entrenched.

Sampling and testing of materials and workmanship

Standardised sampling procedures should be adopted that are in accordance with national standards, where applicable. The frequency of sampling should reflect variability in materials quality and the quality of materials production, mixing, transport and placing processes. Normally, sampling frequency will be prescribed in the specification, but may need to be modified to suit individual circumstances. It should be noted that works performed in-house should generally be subject to the same level of supervision and control as that applied to external contractors.

Resources made available for sampling and testing should be in balance with the frequency of tests required. They should take into account the speed of work, with a view to avoiding unnecessary delay to subsequent operations.

Contractors should be notified immediately of materials not meeting specification requirements. Careful consideration should be given to the reasons for non-compliance before taking action. Possible actions include re-working the material, removing the material, reducing payment or obtaining a further performance bond from the contractor.

9.3.3 Quality management

However, with the move towards functional specifications, the onus is on the contractor to employ quality management systems to assure quality (see Chapter 1). The client's role then becomes one of checking only on a sample basis that functional requirements have been met. The random nature of the sample helps to ensure that all aspects of the work meet requirements, and not just those that are expected to be checked. The use of both functional specifications and the emphasis on quality management encourage a more cost-effective approach by all parties than traditional supervision mechanisms.

9.4 Contractor development

9.4.1 Basis of the approach

In countries with developing or emerging economies, there may be strong developmental arguments for using local construction capacity where this is available. Not only will this capacity usually be less costly to mobilise, but the experience gained will strengthen the industry and help the country to be self-sufficient. There will also often be multiplier effects on other sectors of the economy. Some donors treat local contractors preferentially. For instance, the World Bank will often consider local bids acceptable even if they are 15 per cent more expensive than those from offshore contractors. However, in many of these countries the contracting industry is not well established and, if the efficiency gains from competition are to be realised, then efforts will need to be made to develop the industry.

An approach to development of indigenous contractors has been put forward by the World Bank (Lantran, 1990–1993). This recommends that a development programme has the following components:

- Preliminary study
  - types of works
  - availability and capacities of domestic contractors
  - identification of targets for the development programme.
- Implementing the development programme
  - preparatory training
  - training works
  - assistance to contractors
  - assistance to the road administration.

The recommendations also note that there may be the need to improve institutions and to implement regulatory reforms such as those indicated in Box 9.7. An example of the application of such an approach to contractor development is given by Miles (1996).

9.4.2 Preliminary study

This study should review both the types of works to be undertaken and the existing capacities of the domestic contracting industry. It should then design the development programme by selecting appropriate types and sizes of works to be awarded to selected groups of contractors. The identification of targets then depends on the findings of this study. Realistic targets for the development programme can then be designed in accordance with the current capacities of existing firms, based on reasonable assumptions about contractors' abilities to grow. Both the private contractors and the public administrations may have a limited absorptive capacity: attempting to grow too quickly can result in wasted resources and possible bankruptcy, with consequent effects on the quality and delivery of the product.

9.4.3 *Implementing the development programme*

Preparatory training is the first step in implementing the development programme. This can contain the following:

- Seminars for contractors' staff to update them on managerial techniques.
- Training works, supported by technical assistance to the contractors.
- Other types of assistance to contractors, such as with
  - procurement of equipment
  - access to foreign exchange.
- Development of a supervision and control capability within the road administration.

9.5 Contracts for in-house works

9.5.1 *Improvement of existing operations*

It was noted earlier, that where in-house operations are efficient, there is little evidence of savings by using private contractors. Improvement of the efficiency of existing organisations that operate poorly should, therefore, be an approach that should be considered. The methods of institutional development described in Chapter 2 are relevant to this. The case for making better use of the public sector has been put forward (Madelin, 1993, 1994b), where it is noted that savings can be made by giving a clarity of purpose and a focus for management, and by revitalising the labour force, as in Box 9.10. In order to give in-house operations the chance to improve, there is a need to:

- Separate the client and supplier roles to increase the focus
- Let managers manage
- Apply specifications and use contract procedures
- Publish profit and loss accounts
- Undertake independent audit of procedures and accounts.

The option of the works department of a road administration operating as a contractor bidding against conventional contractors for work has been introduced successfully, for example, in Sweden and in the United Kingdom (Madelin, 1994a). This removes the monopoly of the works department and makes price-fixing between contractors easier to detect.

> **Box 9.10**   *Public sector commercialisation in the United Kingdom*
>
> The early UK experience in improving the efficiency of road maintenance organisations has been described in a World Bank report (Cox, 1987). This examined the effect of the *Direct Labour Organisation (DLO) Legislation* of 1981. This required all local authority works organisations to operate on a commercial basis as contractors to their client authorities. The experience of the DLO legislation suggested that the introduction of a competitive element into operations had resulted in:
>
> - Improvements in the cost-effectiveness of bonus incentive schemes
> - Improved attitudes of employees
> - Gains in operational efficiency
> - Considerable savings in cost
> - No reduction in quality
>
> A more recent example has been quoted for the commercialisation of activities in Shropshire County (Madelin, 1994a,b), where a 15–20 per cent improvement in productivity was obtained in operations by in-house units, and cost savings were achieved, as shown below.
>
> | | |
> |---|---:|
> | Tenders won by direct in-house unit | £4 216 000 |
> | Cost of next lowest contractor | £4 829 999 |
> | Saving | £604 000 |
> | Profit by in-house unit | £100 000 |
> | Estimated effect of competition on private sector prices | £300 000 |
> | Total saving | £1 000 000 |
> | Saving represents 10 per cent of the maintenance budget | |
>
> *Note*:
> Revitalised in-house unit was achieved by:
>
> - Separation of client and supplier functions
> - Unit operating like a private contractor, but only allowed to operate in Shropshire County Council
>
> However, this example noted that there is a need to ensure that competition between public and private sectors is fair. There is also a need to achieve a balance between competition and stability.

9.5.2 *Competitive procurement of activities*

Improved effectiveness and efficiency can be achieved through increased commercialisation of activities leading, ultimately, to contracting out of functions, where this is feasible and appropriate. Commercialisation was discussed in Chapter 2. To achieve this, a series of steps can be taken in the short and medium terms to move the organisation toward this goal.

Referring back to the concepts discussed in Chapter 1, as the management process moves from planning through to operations, the functions being undertaken become easier to contract out. Letting contracts for the planning function is relatively difficult; whereas contracting out the management of operations is relatively easy. It is, therefore, necessary to prioritise the actions that need to be taken to reflect this. Arguably, the following general order of priorities for moving towards competitive procurement can be considered.

Priority of works categories
1. Development works
2. Periodic maintenance
3. Routine maintenance
4. Special works.

Priority of commercialisation measures
1. Functional separation of management from execution (the 'client' function from the 'supplier' function)
2. Other commercial management actions (as described in Chapter 2)
3. Involvement of stakeholders
4. Introducing supplier functions to competition
5. Contracting out the supplier functions
6. Contracting out the client functions.

Ideally, the supplier organisations should be subject to an increasing level of competition. A certain amount of work could be guaranteed by the client organisation, and this could include routine maintenance, emergency and winter maintenance works. The suppliers could bid for contracts from the client organisation, for at least some of the periodic maintenance work and for all of the development work. During this period, it is important that the interests of both the client and the supplier organisations are protected. If the suppliers fail to generate sufficient income to survive, then the client organisation may lose the ability to achieve its periodic maintenance targets. Another situation could arise if the suppliers find it more attractive commercially to work for other clients, again resulting in periodic maintenance targets not being met.

Experience suggests that the best way of increasing competition over time is to set a threshold for the value of work above which all contracts must be subject to competitive tender. Over time, the value of this threshold may be modified. The works unit would, therefore, be guaranteed all periodic maintenance contracts from the client whose value falls below this threshold. This provides some degree of protection for both the client and for the works units during the transition period. Ultimately, all works would be subject to competitive tender, and the works units would then be in a position of relying totally on commercial principles for survival. A medium-term objective could be for all periodic maintenance work to be subject to competitive tender.

Where existing in-house operations are eventually to be privatised, it is likely that they will have good technical expertise, but may lack business skills. Their development can be assisted by:

- Entering into association with well-established businesses.
- Starting with activities requiring low capital investment, and then focusing training programmes on the development of managerial and technical skills.
- Progressive transformation into a private firm through a medium-term process of expanding autonomy.
- Involvement of an international contracting firm in the management of the development process though sub-contracting, joint ventures, or through technical assistance arrangements.

Many countries around the world carry out all periodic maintenance and development works by contract (Madelin, 1994b), and the emphasis in these is now on commercialising routine maintenance and client activities. Issues to be considered for commercialisation were discussed in sub-sections 9.1.4 and 9.1.5.

The pace at which this process can proceed will depend on the objectives of the government and the road administration in particular situations. It will also depend on the management functions and activities, and a summary of the privatisation potential

Table 9.2 *Potential for contracting out different functions and activities*

Functions and activities	Typical location and responsibility	Potential for competitive procurement
Planning (policy)	Central government (Ministry of Transport)	No
Planning (network)	Headquarters of roads administration	No[1]
Programming (including budgeting)	Headquarters of roads administration	No[1]
Preparation (including design):	Regional or district roads administration	
• Routine		Yes (long term)
• Periodic		Yes (medium term)
• Special		Yes (long term)
• Development		Yes (short term)
Operations (routine maintenance):	Regional or district roads administration	
• Supervision (including testing)		Yes (long term)
• Works		Yes (long term)
• Equipment management		Yes (medium term)
• Technical audit		Yes (medium term)
Operations (periodic maintenance):	Regional or district roads administration	
• Supervision (including testing)		Yes (medium term)
• Works		Yes (medium term)
• Equipment management		Yes (medium term)
• Technical audit		Yes (medium term)
Operations (special works):	Regional or district roads administration	
• Supervision (including testing)		Yes (long term)
• Works		Yes (long term)
• Equipment management		Yes (medium term)
• Technical audit		Yes (medium term)

Operations (development):	Regional or district roads	
• Supervision (including testing)	administration	Yes (short term)
• Works		Yes (short term)
• Equipment management		Yes (medium term)
• Technical audit		Yes (medium term)
Other functions and activities:		
• Research	Headquarters organisation	Yes (long term)
• Training	All organisations	Yes (short term)
• Administration, finance and personnel	All organisations	n/a

Note:
(1) Data collection and management information systems used in conjunction with these functions and activities have potential for being privatised.

for each of these is given in Table 9.2.

9.6 General observations

Whereas it is relatively straightforward to execute development and periodic maintenance works by contract, the use of this approach for routine maintenance and special works is more problematic. In all cases, adequate contract management skills are required.

To achieve contract-based works, an interested and skilled domestic contract industry is needed. If this is not the case, efforts will need to be made to develop these skills in the industry either directly, or by the commercialisation or privatisation of public sector organisations. To be successful, contracting must also take place within an appropriate legal, financial and institutional framework that is equitable to both the client and the contracting organisations. Continuity of use is essential if the contracting industry is to be successful, particularly where the market for work varies a great deal from year to year. Long-term competition is also important, if efficiency of operations is to be ensured. The creation of formal or informal cartels should also be prevented, and the possibility of these forming should be guarded against continually. Operating a small in-house operation may prove to be an effective counter-balance to this problem.

The advantages and disadvantages of contract maintenance are summarised in Box 9.11.

Box 9.11   **Summary of advantages and disadvantages of contract maintenance**

Complete operation by in-house unit

Advantages
- The road administration controls the complete process

Disadvantages
- Low flexibility
- High fixed costs
- Irregular utilisation of resources
- Non-specialised equipment
- Few incentives to improve effectiveness and efficiency

Renting or contracting equipment

Advantages
- The road administration controls the management of works
- Peaks of work are cut
- Lower fixed costs
- Special equipment can be rented

Disadvantages
- Requires supervision
- Needs co-ordination with own workforce
- Few incentives for innovation

Contract routine maintenance

Advantages
- Easy procurement
- Budgets are easy to administer
- Limited need for supervision

Disadvantages
- Difficulties describing contractors' responsibilities
- Risk of creating a monopoly supplier

Unit price contracts for routine works

Advantages
- Small risk for the contractor

Disadvantages
- Need for supervision
- Limited incentives for innovation
- Difficulties in predicting total cost

Cost-plus contracts for routine works

Advantages
- Extremely small risk for contractor

Disadvantages
- Need for extensive supervision
- Few incentives for innovation
- Very difficult to predict total cost

Complete competition for all maintenance works using procedural specifications

Advantages
- Small client organisation

Disadvantages
- Poor flexibility
- Risk of monopoly situation
- Few incentives for innovation for the contractor

Complete competition for all maintenance works using functional specifications

Advantages
- Small client organisation
- Incentives for innovation

Disadvantages
- Poor flexibility
- Risk of monopoly situation
- Very difficult to define functional demands

Appendix A: Management Summary

Functions Management	Project	Spatial coverage	Time scales	Staff concerned	Definition of works	Component of policy framework
Planning	Identification	Network-wide	Long term (strategic)	Senior management and policy level	Works category	Mission statement
Programming	Feasibility	Network to sub-network	Medium term (tactical)	Middle-level professional	Works type	Objectives
Preparation	Design and commitment	Section of project (or scheme)	Budget year	Engineer/ technician	Activity	Standards and intervention levels
Operations	Implementation	Sub-section	Immediate/very short term	Technician/works supervisors	Task	
		(Chapter 1)	(Chapter 1)	(Chapter 1)	(Chapter 1)	(Chapter 2)

Contd.

Functions Management	Project	Environmental assessment method	Cost estimating method	Data detail	Computer processing	Contracting out
Planning	Identification	Screening	Global	IQL-IV	Automatic	Relatively difficult
Programming	Feasibility	Preliminary environmental assessment	Unit rate	IQL-III/IV		
Preparation	Design and commitment	Environmental impact assessment		IQL-II/III		
Operations	Implementation	(Review and monitoring)	Operational	IQL-I/II	Interactive	Relatively easy
		(Chapter 2)	(Chapter 4)	(Chapter 5)	(Chapter 5)	(Chapter 9)

References

AASHTO (1987). *AASHTO guidelines for value engineering*. Washington DC: American Association of State Highway and Transportation Officials.

Abaynayaka, S. W. et al. (1977). Prediction of road construction and vehicle operating costs in developing countries. *Proceedings of the Institution of Civil Engineers*, **62** (Part 1), 419–446.

Adair, J. (1983). *Effective leadership*. London: Gower.

Adler, H.A. (1987). Economic appraisal of transport projects: a manual with case studies. *EDI Series in Economic Development*. Baltimore: Johns Hopkins for the World Bank, Revised and expanded edition.

Ahmed, N. U. (1983). An analytical decision model for resource allocation in highway maintenance management. *Transportation Research*, **17A** (2).

Allport, R. J. et al. (1986). The use of scenario techniques to formulate transport strategy for an urban area. In: PTRC – *Transportation Planning Methods, Proc. of Seminar M, PTRC Summer Annual Meeting, University of Sussex, 14–17 July 1978*. London: PTRC Education and Research Services, 229–240.

Amekudzi, A. A. and Attoh-Akine, N. O. (1996). Institutional issues in implementation of pavement management systems by local agencies. *Transportation Research Record 1524*. Washington DC: National Academy Press, 10–15.

American Society of Civil Engineers (1992). *Local low volume roads and streets*. Washington DC: Federal Highway Administration.

Archondo-Callao, R. (1994). HDM Manager Version 3.0. *Transportation Division, Transportation, Water & Urban Development Department*. Washington DC: The World Bank.

Archondo-Callao, R. and Purohit, R. (1989). HDM-PC: user's guide. *The Highway Design and Maintenance Standards Series*. Washington DC: The World Bank.

Atkinson, K. (Ed.) (1997). *Highway maintenance handbook*. London: Thomas Telford, second edition.

Audit Commission (1988). *Improving highways maintenance: a management handbook*. London: H. M. Stationery Office.

Balcerac de Richecour, A. and Heggie, I. G. (1995). African road funds: what works and why? *Sub-Saharan Africa Transport Program SSATP Working Paper No. 14*. Washington DC: The World Bank.

Barber, J. W. (Ed.) (1968). *Industrial training handbook*. London: Iliffe.

Bellman, R. and Dreyfus, S. E. (1962). *Applied dynamic programming*. Princeton NJ: Princeton University Press.

Betz, M. (1966). Interpretation of costs. *ASCE Journal of Highway Division, No. HW2*, **92** (Oct).

Boddy, J. E. (1988). Motorways and trunk road management by consulting engineers and the use of term maintenance contracts. In: *IHT – National Workshop on Reconstruction for Tomorrow's Traffic, Leamington Spa, 26 April 1988*. London: Institution of Highways and Transportation, 121–147.

Bolton, A. (1997). Partnering for progress. *New Civil Engineer*, 8 May 1997, 45–46.

Bridger, G. A. and Winpenny, J. T. (1983). *Planning development projects: a practical guide to the choice and appraisal of public sector investments*. London: H. M. Stationery Office for the Overseas Development Administration.

Britton, C. R. (1991). Highway and pavement management systems – structured approaches to implementation. *Municipal Engineer, June 1991*, **8**, 129–140.

Britton, C. (1994). Computerised road management: the appliance of computer science. *Highways*, **62** (No. 4). London: Thomas Telford, 12–13.

Brooks, D. M. *et al*. (1989). Priorities in improving road maintenance overseas: a check-list for project assessment. *Proceedings of the Institution of Civil Engineers, Part 1*, **86** (Dec), 1129–1141.
Butt, A. A. *et al*. (1994). Application of Markov process to pavement management systems at network level. In: TRB – *Third International Conference on Managing Pavements, San Antonio, Texas, 22–26 May 1994, Proceedings Volume 2*. Washington DC: National Academy Press, 159–172.
Commission of the European Communities (1993). *Project cycle management*. Brussels: Commission of the European Communities.
Cox, B. E. (1987). Evaluation of incentives for efficiency in road maintenance organisation: the UK experience. *Transportation Issues Series Report No. TRP2*. Washington DC: The World Bank.
Croney, D. (1977). *The design and performance of road pavements*. London: H. M. Stationery Office.
Cundill, M. A. and Withnall, S. J. (1995). Road transport investment model RTIM3. In: *Sixth International Conference on Low Volume Roads, Conference Proceedings 6, Volume 1*. Washington DC: National Academy Press, 187–192.
Deighton, R. A. and Blake, D. G. (1994). Improvements to Utah's location referencing system to allow data integration. In: TRB – *Third International Conference on Managing Pavements, San Antonio, Texas, 22–26 May 1994, Proceedings Volume 1*. Washington DC: National Academy Press, 97–107.
Department of Transport (1988). *Routine maintenance management system: highway RMMS surveys – survey procedure manual*. London: Department of Transport.
Department of Transport (1996). *Transport Report 1996* [Cmd 3206]. London: H. M. Stationery Office.
Duffell, J. R. and Pan, J. K. (1996). Minor road deterioration: causes, consequences and maintenance options. *Proceedings of the Institution of Civil Engineers, Transportation*, **117** (Nov), 278–290.
(The) Economist (1996). Taming the beast: a survey on living with the car. *The Economist*, 22 June 1996.
EDI and ECA (1991). The road maintenance initiative. Volume 2: readings and case studies. *Sub-Saharan Africa Transport Program*. Washington DC: The World Bank.
Ellis, C. I. (1975). Risk and the pavement design decision in developing countries. *TRRL Laboratory Report 667*. Crowthorne: Transport and Road Research Laboratory.
Evdorides, H. T. and Snaith, M. S. (1996). A knowledge-based analysis process for road pavement condition assessment. *Proceedings of the Institution of Civil Engineers, Transportation*, **117** (Aug), 202–210.
Farrell, S. (Ed.) (1994). Financing transport infrastructure. *PTRC Perspectives No. 3*. London: PTRC Education and Research Services.
Field, P. C. and Layton, R. D. (1995). Scheduling road maintenance activities with project management software. In: TRB – *Sixth International Conference on Low-Volume Roads, Minneapolis, 25–29 June 1995, Volume 1*. Washington DC: National Academy Press, 257–261.
Freeman-Bell, G. and Balkwill, J. (1993). *Management in engineering: principles and practice*. New York: Prentice Hall.
Gichaga, F. J. and Parker, N. A. (1988). *Essentials of highway engineering*. London: Macmillan.
Haas, R. and Hudson, W. R. (1996). Defining and serving clients for pavements. *Transportation Research Record 1524*. Washington DC: National Academy Press, 1–9.
Haas, R. *et al*. (1994). *Modern pavement management*. Malabar Fa: Krieger.
Hambly, E. C. and Hambly, E. A. (1994). Risk evaluation and realism. *Proceedings Institution of Civil Engineers, Civil Engineering*, **102** (May), 64–71.
Hamilton, M. J. (1996). Privately financed road infrastructure: a concession company's point of view. *SSATP Working Paper No. 26*. Washington DC: The World Bank.
Handy, C. B. (1993). *Understanding organisations*. London: Penguin, third edition.

Harral, C. G. et al. (1979). The highway design and maintenance standards model (HDM): model structure, empirical foundations and applications. *PTRC Summer Annual Meeting, University of Warwick, 13–16 July 1979*. London: PTRC Education and Research Services.

Harral, C. G. et al. (1986). An appraisal of highway maintenance by contract in developing countries. *World Bank Transportation Issues Series Report No. TRP1*. Washington DC: The World Bank.

Hau, T. D. (1992a). Congestion charging mechanisms for roads: an evaluation of current practice. *Policy Research Working Paper WPS 1071*. Washington DC: The World Bank.

Hau, T. D. (1992b). Economic fundamentals of road pricing: a diagrammatic analysis. *Transport Policy Working Papers WPS 1070*. Washington DC: The World Bank.

Heath, W. and Robinson, R. (1980). Review of published research into the formation of corrugations on unpaved roads. *TRRL Supplementary Report 610*. Crowthorne: Transport and Road Research Laboratory.

Heggie, I. G. (1991a). Designing major policy reform: lessons from the transport sector. *World Bank Discussion Papers 115*. Washington DC: The World Bank.

Heggie, I. G. (1991b). Improving management and charging policy for roads: an agenda for reform. *Infrastructure and Urban Development Department Report INU92*. Washington DC: The World Bank.

Heggie, I. G. (1995). Management and financing of roads: an agenda for reform. *World Bank Technical Paper No. 275*. Washington DC: The World Bank.

Hellendoorn, H. (1997). Fuzziness clears the way ahead! *Highways & Transportation*, April 1997, 14–15.

Hindson, J. (edited and revised by Howe, J. and Hathway, G.) (1983). *Earth roads: a practical guide to earth road construction and maintenance*. London: Intermediate Technology Publications.

Howe, J. D. G. F. (1972). A review of rural traffic-counting methods in developing countries. *RRL Report LR427*. Crowthorne: Road Research Laboratory.

Howe, J. (1996). Transport for the poor, or poor transport? *IHE Working Paper IP-12*. Delft: International Institute for Infrastructural, Hydraulic and Environmental Engineering.

Humplick, F. and Paterson, W. D. O. (1994). A framework of performance indicators for managing road infrastructure and pavements. In: TRB – *Third International Conference on Managing Pavements, San Antonio, Texas, 22–26 May 1994, Proceedings Volume 1*. Washington DC: National Academy Press, 123–133.

IHT (1990). *Highway safety guidelines: accident prevention and reduction*. London: Institution of Highways and Transportation, International edition.

Institution of Civil Engineers (1996a). *The engineering and construction contract*. London: Thomas Telford, second edition.

Institution of Civil Engineers (1996b). *Which way roads?* London: Thomas Telford.

ISO (1987). Quality systems. *ISO 9001–1987*. Milton Keynes: British Standards Institution.

Israel, A. (1987). *Institutional development: incentives to performance*. Baltimore: Johns Hopkins for the World Bank.

Jackson, N. C. et al. (1996). Development of pavement performance curves for individual defect indexes in South Dakota based on expert opinion. *Transportation Research Record 1524*. Washington DC: National Academy Press, 130–136.

Johansen, F. (Ed.) (1989). Earmarking, road funds and toll roads. A World Bank Symposium. *Infrastructure and Urban Development Department Report INU45*. Washington DC: The World Bank.

Jones, T. E. (1984a). Dust emissions from unpaved roads in Kenya. *TRRL Laboratory Report 1110*. Crowthorne: Transport and Road Research Laboratory.

Jones, T. E. (1984b). The Kenya maintenance study on unpaved roads: research on deterioration. *TRRL Laboratory Report 1111*. Crowthorne: Transport and Road Research Laboratory.

Jones, T. E. and Petts, R. C. (1991). Maintenance on minor roads using the lengthman contractor system. In: TRB –*Fifth International Conference on Low-Volume Roads, Raleigh, North*

Carolina, 19–23 May 1991, Transportation Research Record 1291, Volume 1*. Washington DC: Transportation Research Board, National Research Council, 41–52.
Jones, T. E. and Robinson, R. (1986). A study of the cost-effectiveness of grading unpaved roads in developing countries. *Research Report 91*. Crowthorne: Transport and Road Research Laboratory.
Keong, C. H. *et al.* (1997). Conditions for successful privately initiated infrastructure projects. *Proceedings of the Institution of Civil Engineers, Civil Engineering*, **120** (May), 59–65.
Kerali, H. R. and Snaith, M. S. (1992). NETCOM: the TRL visual condition model for road networks. *Contractor Report 321*. Crowthorne: Transport Research Laboratory.
Kerali, H. R. *et al.* (1985). Development of a microcomputer based model for road investment in developing countries. In: CIVILCOMP – *Proceedings of the Second International Conference on Civil and Structural Engineering Computing, London, December 1985*. London: Institution of Civil Engineers.
Kerali, H. R. *et al.* (1991). Data analysis procedures for the determination of long term pavement performance relationships for roads in the United Kingdom. In: Institution of Civil Engineers – *Proceedings of the Conference on the US Strategic Highway Research Programme*. London: Thomas Telford.
Kerali, H. R. *et al.* (1996). Data analysis procedures for long-term pavement performance prediction. *Transportation Research Record 1524*. Washington DC: National Academy Press, 152–159.
Korte, T. (1995). Private road maintenance and construction in Finland. In: TRB – *Sixth International Conference on Low-Volume Roads, Minneapolis, 25–29 June 1995, Conference Proceedings 6, Volume 1*. Washington DC: National Academy Press, 221–224.
Lantran, J.-M. (1990–1993). Contracting out of road maintenance activities, volumes I–IV. *Sub-Saharan Africa Transport Policy Program, Road Maintenance Initiative*. Washington DC: The World Bank.
Lantran, J.-M. and Lebussy, R. (1991). Setting up a plant pool. *Road Maintenance Initiative. Contracting Out Road Maintenance Activities. Volume III*. Washington DC: The World Bank.
Lantran, J.-M. and Morse, C. (1995a). Development of the private sector: framework and competition. In: World Bank, European Union and GTZ – *Proceedings of the Highway Policy Seminar for Countries of the Former Soviet Union, Moscow, 15–19 May 1995*. Washington DC: The World Bank, 169–178.
Lantran, J.-M. and Morse, C. (1995b). Development of the private sector: procurement and contract management. In: World Bank, European Union and GTZ – *Proceedings of the Highway Policy Seminar for Countries of the Former Soviet Union, Moscow, 15–19 May 1995*. Washington DC: The World Bank, 179–189.
Lewis N. C. (1996). Traffic congestion and road pricing. *Proceedings of the Institution of Civil Engineers, Transportation*, **117** (May), 122–135.
Liddle, W. J. (1963). Application of AASHO road test results to the design of flexible pavement structures. In: University of Michigan – *Proceedings of the International Conference on the Structural Design of Asphalt Pavements*. Ann Arbor: University of Michigan, 42–51.
Livneh, M. (1994). Repeatability and reproducibility of manual pavement distress survey methods. In: TRB – *Third International Conference on Managing Pavements, San Antonio, Texas, 22–26 May 1994, Proceedings Volume 2*. Washington DC: National Academy Press, 279–289.
Local Authority Associations (1989). *Highway maintenance: a code of good practice*. London: Association of County Councils.
Madelin, K. B. (1993). The case for in-house staff. *Proceedings of the Institution of Civil Engineers: Municipal Engineering*, **98** (June), 85–88.
Madelin, K. (1994a). Highway maintenance management in Shropshire. *Proceedings of the Institution of Civil Engineers: Transportation*, **105** (May), 97–103.
Madelin, K. (1994b). Maintenance by private contractor or direct labour. *PIARC Roads No. 282 I-1994*. Paris: Permanent International Association of Road Congresses, 61–70.
Madelin, K. B. (1996). Is highway maintenance too important for local authorities? *Proceedings of the Institution of Civil Engineers, Transportation*, **117** (Nov), 302–303.

McCoubrey, W. *et al.* (1995). Performance management of road administrations: United Kingdom. In: PIARC – *XXth World Congress, Montreal 1995*. Paris: Permanent International Association of Road Congresses, 409–428.

Miguel, S. and Condron, J. (1991). Assessment of road maintenance by contract. *Infrastructure and Urban Development Department Technical Paper. Report INU91*. Washington DC: The World Bank.

Miles, D. W. J. (1996). Promoting small contractors in Lesotho: privatization in practice. *Proceedings of the Institution of Civil Engineers, Civil Engineering*, **114** (Aug), 124–129.

Millard, R. S. (1993). Road building in the tropics. *Transport Research Laboratory State-of-the-Art Review 9*. London: H. M. Stationery Office.

Moavenzadeh, F. *et al.* (1972). Investment strategies for developing areas: analytical models for choice of strategies in highway transportation. *MIT Department of Civil Engineering Research Report No. 72-62*. Cambridge MA: Massachusetts Institute of Technology.

Mosheni, A. *et al.* (1994). Case study of benefits achieved from improved management of pavement facilities. In: TRB – *Third International Conference on Managing Pavements, San Antonio, Texas, 22–26 May 1994, Proceedings Volume 2*. Washington DC: National Academy Press, 333–341.

Munro-Faure, L. *et al.* (1993). *Achieving quality standards*. London: Pitman for the Institute of Management.

National Audit Office (1991). *Management of road maintenance*. London: H. M. Stationery Office.

Newbery, D. M. *et al.* (1988). Road transport taxation in developing countries. *World Bank Discussion Papers No. 26*. Washington DC: The World Bank.

Nutt, P. *et al.* (1996). The new trunk road agencies. *Proceedings of the Institution of Civil Engineers, Transportation*, **117** (May), 151–154.

OECD (1984). Environmental impact assessment of roads. *Scientific Expert Group E1*. Paris: Organisation for Economic Co-operation and Development.

OECD (1990). *Road monitoring for maintenance management*. Paris: Organisation for Economic Co-operation and Development.

OECD (1995). *Road maintenance management systems in developing countries*. Paris: Organisation for Economic Co-operation and Development.

Oglesby, C. H. and Hicks, R. G. (1982). *Highway engineering*. New York: John Wiley, fourth edition.

Open University (1987). *PMT 605 Project Management Unit 7 Change Control*. Milton Keynes: The Open University Press.

Overseas Development Administration (1988). *Appraisal of projects in developing countries: a guide for economists*. London: H. M. Stationery Office, third edition.

Parsley, L. L. and Robinson, R. (1982). The TRRL road investment model for developing countries (RTIM2). *TRRL Laboratory Report 1057*. Crowthorne: Transport and Road Research Laboratory.

Paterson, W. D. O. (1987). Road deterioration and maintenance effects: models for planning and management. *The Highway Design and Maintenance Standards Series*. Baltimore: Johns Hopkins for the World Bank.

Paterson, W. D. (1991). Choosing an appropriate information system for road management. In: PIARC – *19th World Congress of PIARC, Marrakesh, September 1991*. Paris: Permanent International Association of Road Congresses.

Paterson, W. D. O. and Robinson, R. (1991). Criteria for evaluating pavement management systems. In: Holt, F. B. and Gramling, W. L. (Eds), *Pavement Management Implementation: ASTM STP1121*. Philadelphia: American Society for Testing and Materials.

Paterson, W. D. O. and Scullion, T. (1990). Information systems for road management: draft guidelines on system design and data issues. *Infrastructure and Urban Development Department Report INU 77*. Washington DC: The World Bank.

Peterson, D. E. (1987). Pavement management practices. *NCHRP Synthesis of Highway Practice 135*. Washington DC: Transportation Research Board.

Phillips, S. J. (1994). Development of United Kingdom pavement management system. In: TRB – *Third International Conference on Managing Pavements, San Antonio, Texas, 22–26 May 1994, Proceedings Volume 1*. Washington DC: National Academy Press, 227–236.

PIARC (1994). *International road maintenance handbook, Volumes 1–4*. Crowthorne: Transport Research Laboratory.

Pontin, C. (1986). *Whitehall: tragedy and farce*. London: Hamish Hamilton.

Pouliquen, L.Y. (1970). Risk analysis in project appraisal. *World Bank Staff Occasional Papers No. Eleven*. Baltimore: Johns Hopkins for the World Bank.

Powell, W. D. et al. (1984). The structural design of bituminous roads. *TRRL Laboratory Report 1132*. Crowthorne: Transport and Road Research Laboratory.

Ramjerdi, F. (1995). *Road pricing and toll financing with examples from Oslo and Stockholm*. Oslo and Stockholm: Institute for Transport Economics, Norway and Royal Institute of Technology, Sweden.

Relf, C. and Thriscutt, S. (1991). Human resource development and management. In: EDI and ECA – *Sub-Saharan Africa Transport Program. The Road Maintenance Initiative. Volume 2: Readings and Case Studies*. Washington DC: World Bank, 178–181.

Ritchie, S. G. et al. (1986). Development of an expert system for pavement rehabilitation decision-making. *Transportation Research Record 1070*. Washington DC: Transportation Research Board, 96–103.

Roberts, A. and Pritchard, V. (1991). In search of excellence – can quality assurance help? *Municipal Engineer*, **8** (Dec), 337–344.

Robinson, D. (1997). Privatisation in partnership – a better option than CCT. *Highways & Transportation*, April 1997, 10–11.

Robinson, R. (1988). A view of road maintenance economics, policy and management in developing countries. *Research Report 145*. Crowthorne: Transport and Road Research Laboratory.

Robinson, R. (1991a). Institutional development in road maintenance organizations. In: EDI and ECA – *Sub-Saharan Africa Transport Program, The Road Maintenance Initiative, Volume 2: Readings and Case Studies*. Washington DC: The World Bank, 168–177.

Robinson, R. (1991b). Maintenance. In: Maguire, D. P. (Ed.) – *Appropriate development for basic needs*. London: Thomas Telford for the Institution of Civil Engineers, 203–218.

Robinson, R. (1993). Life cycle costing of highways. In: Bull, J. W. (Ed.) – *Life cycle costing for construction*. London: Blackie Academic & Professional, 53–85.

Robinson, R. and May, P. H. (1997). Road management systems: guidelines for their specification and selection. *Proceedings of the Institution of Civil Engineers, Transport*, **123** (Feb), 9–16.

Robinson, R. and Phillips, S. J. (1992). The UK pavement management system (UKPMS): optimising treatment priorities. In: IHT – *Alan Brant National Workshop, Royal Spa Centre, Leamington Spa, 14 April 1992*. London: Institution of Highways and Transportation, 95–102.

Robinson, R. and Roberts, P. (1982). The cost-effectiveness of road maintenance. In: UNECA – *Third African Highway Maintenance Conference, Addis Ababa, September 1982*. Addis Ababa: UN Economic Commission for Africa.

Sadek, A. W. et al. (1996). Deterioration prediction modeling of Virginia's interstate highway system. *Transportation Research Record 1524*. Washington DC: National Academy Press, 118–129.

Salter, R. J. (1993). *Highway design and construction*. London: Macmillan, second edition.

Saraf, C. L. and Majidzadeh, K. (1992). Distress prediction models for a network-level pavement management system. *Transportation Research Record 1344*. Washington DC: National Academy Press, 38–48.

Sayers, M. W. et al. (1986). Guidelines for conducting and calibrating road roughness measurements. *World Bank Technical Paper No. 46*. Washington DC: The World Bank.

Schacke, I. and Ertman Larsen, H. J. (1990). *Objectives, achievements and problems in setting up and operating a pavement management system*. Copenhagen: Danish Road Laboratory.

Schliesser, A. and Bull, A. (1993). *Roads: a new approach for road network management and conservation* (LC/L.693). Santiago: UN Economic Commission for Latin America and the Caribbean.
Shahin, M. Y. (1994). *Pavement management for airports, roads, and parking lots*. New York: Chapman & Hall.
Shell (1978). *Shell pavement design manual*. London: Shell International Petroleum Company.
Smith, G. (1995). Financing road expenditures: EBRD perspective. In: World Bank, European Union and GTZ – *Proceedings of the Highway Policy Seminar for Countries of the Former Soviet Union, Moscow, 15–19 May 1995*. Washington DC: The World Bank, 107–109.
Smith, R. B. et al. (1994). Contract road maintenance in Australia. In: TRB – *Third International Conference on Managing Pavements, San Antonio, Texas, 22–26 May 1994, Proceedings Volume 2*. Washington DC: National Academy Press, 113–121.
Smith, R. E. (1994). New approach to defining pavement management implementation steps. In: TRB – *Third International Conference on Managing Pavements, San Antonio, Texas, 22–26 May 1994, Proceedings Volume 2*. Washington DC: National Academy Press, 148–156.
Smith, R. E. and Hall, J. P. (1994). Overview of institutional issues in pavement management implementation and use. In: TRB – *Third International Conference on Managing Pavements, San Antonio, Texas, 22–26 May 1994, Proceedings Volume 2*. Washington DC: National Academy Press, 53–63.
Snaith, M. S. (1985). Pavement condition assessment techniques. *Highways and Transportation, January 1985*. London: Institution of Highways and Transportation, 11–15.
Snaith, M. S. et al. (1994). Knowledge-based systems for maintenance. In: TRB – *Third International Conference on Managing Pavements, San Antonio, Texas, 22–26 May 1994, Proceedings Volume 1*. Washington DC: National Academy Press, 29–36.
Talvitie, A. P. (1996). International experiences in restructuring the road sector. *Training Seminar on Management and Financing of Roads, The World Bank, May 14–15, 1996*.
Thagesen, B. (Ed.) (1996). *Highway and traffic engineering in developing countries*. London: Spon.
Transportation Research Board (1994). *Third International Conference on Managing Pavements, San Antonio, Texas, 22–26 May 1994*. Washington DC: National Academy Press.
TRRL Overseas Unit (1985). Maintenance techniques for district engineers. *Overseas Road Note 2*. Crowthorne: Transport and Road Research Laboratory, second edition.
TRRL Overseas Unit (1987). Maintenance management for district engineers. *Overseas Road Note 1*. Crowthorne: Transport and Road Research Laboratory, second edition.
TRRL Overseas Unit (1988a). A guide to road project appraisal. *Overseas Road Note 5*. Crowthorne: Transport and Road Research Laboratory.
TRRL Overseas Unit (1988b). A guide to bridge inspection and data systems for district engineers, 2 volumes. *Overseas Road Note 7*. Crowthorne: Transport and Road Research Laboratory.
TRRL Overseas Unit (1988c). A guide to geometric design. *Overseas Road Note 6*. Crowthorne: Transport and Road Research Laboratory.
TRRL Overseas Unit (1991). *Towards safer roads in developing countries*. Crowthorne: Transport and Road Research Laboratory.
UMIST Project Management Group (1989). *A guide to cost estimating for overseas construction projects*. London: Overseas Development Administration.
Vaidya, K. G. (1983). Guide to the assessment of rural labour supply for labour-based construction projects. *World Employment Programme Report CTP21*. Geneva: International Labour Office.
Van der Tak, H. G. and Ray, A. (1971). The economic benefits of road transport projects. *World Bank Staff Occasional Papers No. 13*. Baltimore: Johns Hopkins for the World Bank.
Wang, K. C. P. et al. (1994). Design of project selection procedure based on expert systems and network optimization. In: TRB – *Third International Conference on Managing Pave-

ments, San Antonio, Texas, 22–26 May 1994, Proceedings Volume 2. Washington DC: National Academy Press, 173–183.

Watanatada, T. *et al.* (1987). The Highway Design and Maintenance Standards Model. Volume 1: description of the HDM-III model. *The Highway Design and Maintenance Standards Series*. Baltimore: Johns Hopkins for the World Bank.

World Bank (1988). Road deterioration in developing countries: causes and remedies. *A World Bank Policy Study*. Washington DC: The World Bank.

World Bank (1994). Roads and the environment: a handbook. *Report TWU 13*. Washington DC: The World Bank.

Zimmermann, K. A. and Darter, M. I. (1994). Using innovative management techniques in implementing pavement management systems. In: TRB – *Third International Conference on Managing Pavements, San Antonio, Texas, 22–26 May 1994, Proceedings Volume 2*. Washington DC: National Academy Press, 139–147.

Index

Accidents 4, 10, 94
Accounting systems
   accrual 99
   cash 98
Activities (definition) 2
Advance payment guarantee 269
Agency agreements 254
AGETIP 254
Asset value 1
Attributes 133
Axle loading 8, 141

Baseline: do nothing or minimum 113
Benefits
   economic, road user 4, 8, 86
   network and project level 95
Bond: bid and performance 268
Brooks pyramid 56
BSM 192
Budget: allocation 64–9
Budget constraint 12, 116, 196–210
Budget head 66
Budgeting 65–9
Business management 5

Capital expenditure 66
CHART 192
Client function 53, 253–6, 273
Commercial models 190
Commercialisation 55, 274–6
Competition 50, 52, 248, 260
Competitive procurement 274–6
Concession financing 84
Condition-responsive 161, 163
Congestion 76
Construction 5
Construction costing 99–112
Contract packaging 186, 187
Contract types 264
Contracting out 52
Contractual risks 88
Cordon pricing 79
Corrugations 180
Cost
   average annual 197
   present value of (PVC) 113–14
   survey 225

Cost control 245
Cost–benefit analysis 112–16
Cost-effectiveness 196
Costing: global, unit rate,
   worker hours 99
Customers 17, 34
Cyclic maintenance, works 2, 36, 222

Data acquisition costs 122
Data collection 126–49
Database management software 223
Decentralisation 46, 48
Decision tree 176–7
Decision-support system 153, 188
Defect lengths 165
Defects
   gravel roads 180
   paved roads 142
De-icing 232
Depreciation 245
Deterioration 8, 164, 210–18
Development
   career 26
   economic and social 3
   institutional, organisational 56–60
   road (definition) 3
Diagnostic tool 40
Disbursement mechanisms 68
Discounted costs 9, 10, 113
Dust 181–2
Dynamic sectioning,
   segmentation 132, 165

Economic appraisal, feasibility 12, 112
Economic boundary 202–7
Effective gradient 202
Effectiveness (definition) 50
Efficiency (definition) 50
Efficiency frontier *see* Economic
   boundary
Emergency works (definition) 3
Entity 133
Environmental appraisal, assessment 61
Environmental costs 81
Environmental degradation 4
Environmental impact assessment 62
Environmental pollution 77–8

Equipment management 241
Expert system 174
External factors 26
Externalisation 255

Feature (definition) 133
Fuel levy 71
Fund allocation 64–9
Funds: source 47

Geographic information
  system (GIS) 131
Global positioning system (GPS) 131
Grading 183

HCM 119
HDM-III/HDM-4 119
HERS 208
Hierarchy: road 38
Hierarchy of management issues 57

Impacts 2, 3
Incremental net present value/cost 203
Inflation 102, 113
Information groups 122, 123
Information quality levels (IQL) 126
Information system 41, 153
Inspection 224
Institutional factors 26
Intermediate technology 222
Internal rate of return 115
Intervention levels 32, 36–8, 168, 191

Labour based, labour intensive 222
Lane rental contracts 250
Lengthmen 252
Level of service 4
Licence fee 72
Life cycle cost 9, 66, 117, 120

Management cycle 14, 121
MARCH 192
Materials sampling and testing 271
Mission statement 32–5
Mutual exclusivity 114

Net present value (NPV) 115
Net present value/cost ratio 116, 196
NETCOM 96
New Engineering Contract (NEC) 266
Nodes 131

Objective function 200
Objectives 32–6

Operations (definition) 13
Operations
  effectiveness and efficiency
    50, 246, 277–8
  equipment 241–5
  impacts 2, 93–7
  toll 79
Opportunity cost 112
Optimisation 200
Organisational culture 43–4
Organisational development 44
Organisational reform 55
Organisational structure 46
Ownership 29–30

Parking charges 77
Partnering 266
Pavement texture and friction
  143, 146, 148
Performance curves 197
Performance indicators 40–2
Performance standards 103, 106
Periodic works (definition) 2
Planning (definition) 12
Policy document 32
Policy framework 31–43
Policy reform 58
Preparation (definition) 12
Present serviceability index (PSI) 191
Present value of cost (PVC) 113–14
Preventive works (definition) 2
Price indices 100
Priority index 191
Private finance 83
Privatisation 52, 260
Probability distribution 211
Process (definition) 4
Programming (definition) 12
Project (definition) 5
Project cycle 110
Project plan 270
Project preparation 124
Project screening 61

Qualification: pre-, post- 267–8
Quality assurance (QA) 17, 21
Quality control 21

Reactive works (definition) 2
Recurrent expenditure 66
Rental rates for equipment 104, 105
Resource allocation 41
Resource costs 113
Risk analysis, risk management 22–3

Index   291

RMMS   226
Road administration, road agency,
    road authority   49–50, 53
Road administration cost   4, 9, 190
Road fund   70, 74
Road pricing   80
Road tariff   70
Road user charges   71, 73
Road user cost   4, 93, 190
Roads: proclaimed, unproclaimed   30
Roads board   28
Rolling programme of works   206
Roughness   143, 145, 146, 180
Routine works (definition)   2
RTIM   119

Safety   10, 230–5
Sampling rates   149, 151
Scenario analysis   22
Scheduled works   161
Scheduling of works   226
Sensitivity analysis   22
Shadow price   114
Shadow toll   88
Short-run marginal cost   71
Snow clearance   230–2
Spare parts management   245
Special works (definition)   3
Specification: end product, functional,
    method, procedural   264–5

Specificity   50
Staff requirements   47
Stakeholders   29, 69
Standards: road   32, 36–8, 168, 191
Strategic planning   see Planning
Structural capacity   143, 145, 147
Supplier function   53
Surface distress   143, 145, 147

Target level   37
Technical factors   25
Thermal mapping   235
Thresholds   37
Tolls   79
Total annualised system cost (TASC)   209
Total quality management (TQM)   21
Traffic flow   136
Traffic growth   140
Training   59–60
Transition matrix   212

User model   190
Utility charge   70

Value engineering   23
Vehicle operating cost   8, 9

Warning level   37
Works activities   2
Works scheduling   226

Learning Resources
Centre